ALTERNATIVES TO THE **Indian Point Energy Center** FOR MEETING
NEW YORK ELECTRIC POWER NEEDS

Committee on Alternatives to Indian Point for Meeting Energy Needs

Board on Energy and Environmental Systems
Division on Engineering and Physical Sciences

NATIONAL RESEARCH COUNCIL
OF THE NATIONAL ACADEMIES

THE NATIONAL ACADEMIES PRESS
Washington, D.C.
www.nap.edu

THE NATIONAL ACADEMIES PRESS • 500 Fifth Street, N.W. • Washington, DC 20001

NOTICE: The project that is the subject of this report was approved by the Governing Board of the National Research Council, whose members are drawn from the councils of the National Academy of Sciences, the National Academy of Engineering, and the Institute of Medicine. The members of the committee responsible for the report were chosen for their special competences and with regard for appropriate balance.

This report and the study on which it is based were supported by Contract No. DE-AT01-04TD45037 (Task Order No. 6) from the U.S. Department of Energy. Any opinions, findings, conclusions, or recommendations expressed in this publication are those of the author(s) and do not necessarily reflect the view of the organizations or agencies that provided support for the project.

International Standard Book Number: 0-309-10172-7

Cover: The transmission network links generating plants, including Indian Point, with demand centers in all parts of New York State. Map courtesy of the New York State Independent System Operator. Indian Point Energy Center image courtesy of Entergy Corporation.

Available in limited supply from:
Board on Energy and Environmental
  Systems
National Research Council
500 Fifth Street, N.W.
Keck W934
Washington, DC 20001
202-334-3344

Additional copies available for sale from:
The National Academies Press
500 Fifth Street, N.W.
Lockbox 285
Washington, DC 20055
800-624-6242 or 202-334-3313
(in the Washington metropolitan area)
http://www.nap.edu

Copyright 2006 by the National Academy of Sciences. All rights reserved.

Printed in the United States of America

THE NATIONAL ACADEMIES
Advisers to the Nation on Science, Engineering, and Medicine

The **National Academy of Sciences** is a private, nonprofit, self-perpetuating society of distinguished scholars engaged in scientific and engineering research, dedicated to the furtherance of science and technology and to their use for the general welfare. Upon the authority of the charter granted to it by the Congress in 1863, the Academy has a mandate that requires it to advise the federal government on scientific and technical matters. Dr. Ralph J. Cicerone is president of the National Academy of Sciences.

The **National Academy of Engineering** was established in 1964, under the charter of the National Academy of Sciences, as a parallel organization of outstanding engineers. It is autonomous in its administration and in the selection of its members, sharing with the National Academy of Sciences the responsibility for advising the federal government. The National Academy of Engineering also sponsors engineering programs aimed at meeting national needs, encourages education and research, and recognizes the superior achievements of engineers. Dr. Wm. A. Wulf is president of the National Academy of Engineering.

The **Institute of Medicine** was established in 1970 by the National Academy of Sciences to secure the services of eminent members of appropriate professions in the examination of policy matters pertaining to the health of the public. The Institute acts under the responsibility given to the National Academy of Sciences by its congressional charter to be an adviser to the federal government and, upon its own initiative, to identify issues of medical care, research, and education. Dr. Harvey V. Fineberg is president of the Institute of Medicine.

The **National Research Council** was organized by the National Academy of Sciences in 1916 to associate the broad community of science and technology with the Academy's purposes of furthering knowledge and advising the federal government. Functioning in accordance with general policies determined by the Academy, the Council has become the principal operating agency of both the National Academy of Sciences and the National Academy of Engineering in providing services to the government, the public, and the scientific and engineering communities. The Council is administered jointly by both Academies and the Institute of Medicine. Dr. Ralph J. Cicerone and Dr. Wm. A. Wulf are chair and vice chair, respectively, of the National Research Council.

www.national-academies.org

COMMITTEE ON ALTERNATIVES TO INDIAN POINT FOR MEETING ENERGY NEEDS

LAWRENCE T. PAPAY, NAE,[1] Consultant, *Chair*
DAN E. ARVIZU, National Renewable Energy Laboratory
JAN BEYEA, Consulting in the Public Interest
PETER BRADFORD, Bradford Brook Associates, Ltd.
MARILYN A. BROWN, Oak Ridge National Laboratory
ALEXANDER E. FARRELL, University of California, Berkeley
SAMUEL M. FLEMING, Consultant
GEORGE M. HIDY, Envair/Aerochem
JAMES R. KATZER, NAE, Consultant
PARKER D. MATHUSA, New York State Energy Research and Development Authority
TIMOTHY MOUNT, Cornell University
FRANCIS J. MURRAY, JR., Consultant
D. LOUIS PEOPLES, Nyack Management Company, Ltd.
WILLIAM F. QUINN, Argos Utilities LLC
DAN W. REICHER, New Energy Capital Corporation
JAMES S. THORP, NAE, Virginia Polytechnic Institute and State University
JOHN A. TILLINGHAST, NAE, Tillinghast Technology Interests, Inc.

Project Staff

Board on Energy and Environmental Systems (BEES)

ALAN CRANE, Study Director
DUNCAN BROWN, Senior Program Officer (part time)
JAMES J. ZUCCHETTO, Director, BEES
PANOLA GOLSON, Program Associate

Consultants

General Electric International, Inc.
Optimal Energy, Inc.

---

[1] NAE, National Academy of Engineering.

BOARD ON ENERGY AND ENVIRONMENTAL SYSTEMS

DOUGLAS M. CHAPIN, NAE,[1] MPR Associates, Inc., *Chair*
ROBERT W. FRI, Resources for the Future (senior fellow emeritus), *Vice Chair*
RAKESH AGRAWAL, NAE, Purdue University
ALLEN J. BARD, NAS,[2] University of Texas, Austin
DAVID L. BODDE, Clemson University
PHILIP R. CLARK, NAE, GPU Nuclear Corporation (retired)
MICHAEL L. CORRADINI, NAE, University of Wisconsin, Madison
E. LINN DRAPER, JR., NAE, American Electric Power, Inc. (emeritus)
CHARLES GOODMAN, Southern Company
DAVID G. HAWKINS, Natural Resources Defense Council
MARTHA A. KREBS, California Energy Commission
DAVID K. OWENS, Edison Electric Institute
WILLIAM F. POWERS, NAE, Ford Motor Company (retired)
TONY PROPHET, Carrier Corporation
MICHAEL P. RAMAGE, NAE, ExxonMobil Research and Engineering Company (retired),
MAXINE SAVITZ, NAE, Honeywell, Inc. (retired)
PHILIP R. SHARP, Harvard University
SCOTT W. TINKER, University of Texas, Austin

Staff

JAMES J. ZUCCHETTO, Director
DUNCAN BROWN, Senior Program Officer (part time)
ALAN CRANE, Senior Program Officer
MARTIN OFFUTT, Senior Program Officer
DANA CAINES, Financial Associate
PANOLA GOLSON, Program Associate
JENNIFER BUTLER, Financial Assistant

---

[1] NAE, National Academy of Engineering.
[2] NAS, National Academy of Sciences.

Preface

The Indian Point Energy Center, with two operational nuclear reactors, is in a densely populated region about 40 miles north of midtown Manhattan. On September 11, 2001, one of the hijacked planes flew past the plant on the way to the World Trade Center. This incident heightened concerns that a terrorist attack on the reactors or the spent fuel pools might cause a catastrophic release of radioactivity and led to calls for the plant to be closed.

The Indian Point Energy Center is a vital part of the system supplying electricity to the New York City region. Any significant interruption of power to New York City also could have serious consequences, as shown by the relatively brief blackout that occurred in August 2003. The system delivering power to New York City consumers must be highly reliable, and that depends on having adequate generating capacity available.

This dichotomy led the U.S. Congress to request a study from the National Academies on potential options for replacing the energy services provided by Indian Point. The request, initiated by Representative Nita M. Lowey of New York's 18th District, was directed to the U.S. Department of Energy, which in turn arranged for the study with the National Research Council (NRC) of the National Academies.

The NRC established the Committee on Alternatives to Indian Point for Meeting Energy Needs to conduct the study. Committee members were selected from industry, academia, national laboratories, and other organizations for their expertise on electric power technology and systems and on issues specific to New York. Biographical sketches of the committee members are presented in Appendix A.

The committee was charged with fulfilling the following statement of task:

The National Academies' National Research Council will form a committee to review options for replacing current electric power generation from the Indian Point Energy Center (New York) nuclear facilities with alternative means for meeting electric power demand and associated energy services. The study may include consideration of fossil-fuel-based options (e.g., coal-fired or natural-gas-fired power generation), renewable-energy-based options (e.g., wind, solar, biomass), imports of required electrical energy, and energy efficiency measures, or some combination thereof. The study should include an assessment of the pros and cons of the alternatives to the continued operation of the Indian Point nuclear power plants. The study will not result in the choice of an option but will compare options based on the criteria adopted by the committee.

In 2005, the committee met twice in Washington, D.C., and once in White Plains, New York, to gather information from public sources. The committee was particularly interested in the feasibility of implementing the various options on a scale sufficient to replace the 2,000 megawatts of electric power now produced by Indian Point and to address the resulting economic, environmental, and societal impacts. It procured the services of General Electric International, Inc., to model the New York electric system and how the options would affect reliability. It also contracted with Optimal Energy, Inc., to detail the efficiency improvements that could be made in the New York City area, based on its statewide assessment for the New York State Energy Research and Development Authority. The committee also met twice in closed session to discuss results and progress on this report and held numerous conference calls. Details of the meetings are provided in Appendix B.

The report focuses exclusively on options for replacing current electric power generation and ancillary services from Indian Point. In accordance with the original request, it does not examine the potential for terrorist attacks on Indian Point, nor their probability of success or possible consequences. It makes no recommendations as to whether Indian Point should be closed or how that decision could be implemented. The overriding goal of the study was to evaluate the options that are available to meet electric power demand and to provide the other services required to maintain the reliability of the electric system should a decision be made to close the Indian Point plant.

This report presents the committee's findings. It is the result of a great deal of effort on the part of many highly qualified experts. I greatly appreciate the efforts by the committee members and their enthusiasm, dedication, and insights in conducting this study and preparing the report. The committee operated under the auspices of the NRC Board on Energy and Environmental Systems and is grateful for the able assistance of James Zucchetto, Alan Crane, Panola Golson, and Duncan Brown of the NRC staff.

Lawrence T. Papay, *Chair*
Committee on Alternatives to Indian
Point for Meeting Energy Needs

Acknowledgments

The Committee on Alternatives to Indian Point for Meeting Energy Needs is grateful to the many individuals who contributed their time and effort to the National Academies' National Research Council (NRC) study. The presentations at committee meetings provided valuable information and insight on energy options and constraints in the New York area. The committee thanks the following individuals who provided briefings:

Beth Tritter, Office of Congresswoman Nita M. Lowey,
Philip Overholt, U.S. Department of Energy,
John Kucek, Oak Ridge National Laboratory,
Lawrence Pakenas, New York State Energy Research and Development Authority,
John Plunkett, Optimal Energy, Inc.,
Randall Swisher, American Wind Energy Association,
Harry Vidas, Energy and Environmental Analysis, Inc.,
Philip Fedora, Northeast Power Coordinating Council,
Bill Quinn, Argos Utilities, LLC,
Juanita Haydel, ICF Consulting,
Michael R. Kansler, Entergy Nuclear Northeast,
Steve Mitnick, Conjunction, LLC,
Howard Tarler, New York State Department of Public Service,
The Honorable Andrew J. Spano, Westchester County Executive,
The Honorable Michael Kaplowitz, Westchester County Board of Legislators,
Bruce Biewald, Synapse Energy Economics, Inc.,
Alex Matthiessen, Riverkeeper,
Fred Zalcman, Pace Law School Energy Project,
Garry Brown, New York Independent System Operator,
Michael Forte, Consolidated Edison,
Carl Seligson, Economic and Strategic Consultant,
N.Z. Shilling, GE, and
Paul A. DeCotis, New York State Energy Research and Development Authority.

This report has been reviewed in draft form by individuals chosen for their diverse perspectives and technical expertise, in accordance with procedures approved by the NRC's Report Review Committee. The purpose of the independent review is to provide candid and critical comments that will assist the institution in making its published report as sound as possible and to ensure that the report meets institutional standards for objectivity, evidence, and responsiveness to the study charge. The review comments and draft manuscript remain confidential to protect the integrity of the deliberative process. We wish to thank the following individuals for their review of this report:

David Bodde, Clemson University,
William L. Chameides (NAS), Environmental Defense,
Douglas M. Chapin (NAE), MPR Associates, Inc.,
Michehl R. Gent, Summit Power,
Leonard S. Hyman, RJ Rudden Associates,
Paul Komor, University of Colorado,
Gerald L. Kulcinski, University of Wisconsin,
Harold N. Scherer, Jr. (NAE), Board of Directors, New York Independent System Operator,
Robert J. Thomas, Cornell University,
Harry Vidas, Energy and Environmental Analysis, Inc.,
Carl Weinberg, Weinberg Associates, and
Irvin L. (Jack) White, formerly with Pacific Northwest National Laboratory and New York State Energy Research and Development Authority.

Although the reviewers listed above have provided many constructive comments and suggestions, they were not asked to endorse the conclusions or recommendations, nor did they see the final draft of the report before its release. The review of this report was overseen by George Hornberger (NAE), University of Virginia. Appointed by the National Research Council, he was responsible for making certain that an independent examination of this report was carried out in accordance with institutional procedures and that all review comments were carefully considered. Responsibility for the final content of this report rests entirely with the authoring committee and the institution.

The committee offers special thanks to Mark Sanford, Gene Hinkle, and Gary Jordan at GE Energy and to John Adams and William Lamanna at the New York Independent System Operator for their efforts on the committee's scenario analysis. The committee also benefited from an analysis of energy efficiency opportunities by John Plunkett and Optimal Energy, Inc.

The committee is also very appreciative of the contributions of Erin Hogan, Paul DeCotis, and John Spath of the New York State Energy Research and Development Authority; Benjamin Sovacool of Oak Ridge National Laboratory; and Lynn Billman, Robert Margolis, Brian Parsons, Ralph Overend, Rich Bain, Phil Shepherd, and Walter Short of the National Renewable Energy Laboratory.

Contents

ABSTRACT 1

SUMMARY AND FINDINGS 3

1 INTRODUCTION 8
   Background, 8
     Electricity Supply and Demand, 8
     The Indian Point Energy Center: Description and Role, 14
     Community Concerns, 14
   Criteria for Evaluating Replacement Options, 15
   Conduct of the Study, 16
   Organization of the Report, 17
   References, 17

2 DEMAND-SIDE OPTIONS 18
   Demand Growth in the Indian Point Service Area, 18
   Potential of Demand-Side Options, 20
     Definition of Demand-Side Options and Measures of Potential, 21
     Current Programs Operating in the Indian Point Territory, 23
     The Potential for Additional Energy-Efficiency Improvements, 26
     The Potential for Future Demand Response, 27
     The Potential for Expanded Combined Heat and Power, 29
     The Potential for Expanded Distributed Photovoltaics, 29
     Summary, 30
     Impediments to Demand-Side Programs, 31
   References, 33

3 GENERATION AND TRANSMISSION OPTIONS 35
   Existing Generating Capacity, 35
   Potential New Generating Capacity, 36
     Technologies Considered, 36
     Overall Considerations, 40
   Electrical Transmission, 40
     Existing Transmission, 40
     New Transmission, 41
   Reliability and Reactive Power, 42
     Reliability, 42
     Reactive Power, 43
   References, 43

4 INSTITUTIONAL CONSIDERATIONS AND
   CHANGING IMPACTS   44
   Regulation, Finance, and Reliability, 44
      The New York State Electricity Market, 44
      The Permitting Process with Article X, 50
   Social Concerns, 51
      Environmental Regulation, 51
      Energy Security, 56
      Socioeconomic Factors Including Indirect Costs to the Public, 56
   References, 57

5 ANALYSIS OF OPTIONS FOR MEETING
   ELECTRICAL DEMAND   59
   The NYISO Starting Point, 59
   The Committee's Reference Case, 60
   Replacement Scenarios, 62
   Results of Reliability Analyses, 63
   Operational and Economic Impacts, 66
      Analytical Considerations, 67
      Fuel Diversity: Impact on NYCA Reliance on Natural Gas for
         Generating Electricity, 68
      Projected Impact on the Wholesale Price of Electricity, 69
      Impact on the Annual Variable Cost of Producing Electricity, 71
      Sensitivity to Higher Fuel Prices, 72
   Comparing the Results with Criteria, 73
   References, 74

APPENDIXES[1]

A   Committee Biographical Information   77
B   Presentations and Committee Meetings   82
C   Acronyms   84
D   Supply Technologies   86
    D-1   Cost Estimates for Electric Generation Technologies, 87
    D-2   Zonal Energy and Seasonal Capacity in New York State,
            2004 and 2005, 94
    D-3   Energy Generated in 2003 from Natural Gas Units in
            Zones H Through K, 104
    D-4   Proposed Pipeline Projects in the Northeast of the United States, 105
    D-5   Coal Technologies, 106
    D-6   Generation Technologies—Wind and Biomass, 110
    D-7   Distributed Photovoltaics to Offset Demand for Electricity, 118
E   Paying for Reliability in Deregulated Markets   124
F   Background for the System Reliability and Cost Analysis   144
    F-1   The NYISO Approach, 145
    F-2   Notes on the MARS-MAPS Simulations, 148
G   Demand-Side Measures   169
    G-1   Demand Reduction, 170
    G-2   Estimating the Potential for Energy-Efficiency Improvements, 171
    G-3   Estimating Demand-Response Potential, 175
    G-4   Estimating Photovoltaics for Demand Reduction, 176

---

[1]Appendixes D through G are reproduced on the CD-ROM that contains the full report but are not included in the printed report owing to space limitations.

Tables, Figures, and Boxes

TABLES

2-1 Weather-Normalized Annual Electricity Use, Past and Forecast, in Gigawatt-Hours per Year, for Three New York Regions and Statewide, Selected Years from 1993 Through 2015, 19

2-2 Weather-Normalized Summer Peak Power, Past and Forecast, in Megawatts, for Three New York Regions and Statewide, Selected Years from 1993 Through 2015, 19

2-3 Current Photovoltaic (PV)-Related Policies in New York State, 24

2-4 Committee Estimation of the Potential of Energy-Efficiency Programs in New York Control Area Zones I, J, and K, Selected Years Between 2007 and 2015 (MW), 27

2-5 Committee Estimation of Potential Peak Reduction from Demand-Response Programs in New York Control Area Zones I, J, and K, Selected Years Between 2007 and 2015 (MW), 29

2-6 Committee Estimation of Potential Peak Reduction from Combined Heat and Power in New York Control Area Zones I, J, and K, Selected Years Between 2007 and 2015 (MW), 29

2-7 Committee Estimation of Potential Peak Reduction from Photovoltaics in New York Control Area Zones I, J, and K, Selected Years Between 2007 and 2015, 30

3-1 Approximate (Noncoincident) Summer Peak Load and Capacity in New York State, by Region, 35

3-2 Potential Generating Technologies Considered by the Committee for Replacing Indian Point, 37

3-3 Nominal Transfer Capability Between New York Regions, 41

4-1 Estimated Future Emission Allowance Prices, 54

4-2 Annual Costs for Allowances to Replace Indian Point Generation, Without CO_2 Control (Regional Greenhouse Gas Initiative Baseline Scenario, No CO_2 Control), 55

4-3 Annual Costs for Allowances to Replace Indian Point Generation with CO_2 Control (Regional Greenhouse Gas Initiative Reference Scenario), 55

5-1 NYISO Base Case Peak Load and Known New York Control Area (NYCA) Resources, 60

5-2 Additional Generating Capacity Assumed in Reference Case, 61

5-3   Capacity Additions Assumed for Cases b2 and c2, 64
5-4   Summary of Illustrative Resources Assumed to Maintain NYCA Reliability, 64
5-5   Results of Reliability Analyses, 65
5-6   Benchmark of the Consumption of Natural Gas, Coal, and Oil for 2005 and 2008: Annual Fuel Consumption in Trillion Btu, 69
5-7   Projected Impact on Electrical Generation Based on Natural Gas for 2008 to 2015, with Sensitivity to Fuel Price, 69
5-8   MAPS-Projected Impact on Electricity Wholesale Price, 70
5-9   Projected Impact on Annual Variable Operating Cost, 72

D-1-1   Summary Cost Estimates: Total Cost of Electricity (in 2003 U.S. dollars per kilowatt-hour) for Generating Technologies Examined by the Committee, 87
D-1-2   Cost Components for Electricity Generation Technologies, 88
D-1-3a   Energy Information Administration National Average Cost Estimates (2003 dollars), 89
D-1-3b   Energy Information Administration National Average Cost Estimates (2003 dollars), 90
D-1-4a   Energy Information Administration Regional Cost Estimates (2003 dollars), 91
D-1-4b   Energy Information Administration Regional Cost Estimates (2003 dollars), 92
D-1-5   University of Chicago National Average Cost Estimates (2003 dollars), 92
D-1-6   University of Chicago Regional Cost Estimates for the New York Control Area (2003 dollars), 93
D-1-7   New York City Fuel Prices ($/MMBtu), 93
D-2-1   Summary of Summer and Winter Capacity, Energy Production, and Energy Requirements in the New York Control Area, by Zone, 94
D-2-2   Summer Zonal Capacity, by Fuel, 2004 and 2005, 95
D-2-3   Winter Zonal Capacity, by Fuel, 2004 and 2005, 96
D-2-4   Annual Energy Production, by Fuel, 2004 and 2005, 97
D-2-5   Summary of New York Control Area Generation Facilities' Energy Production by Fuel Type as of January 1, 2005, 98
D-2-6   Summary of New York Control Area Generation Facilities' Winter Capacity, by Fuel Type, as of January 1, 2005, 99
D-2-7   Summary of New York Control Area Generation Facilities' Summer Capacity, by Fuel Type, as of January 1, 2005, 100
D-2-8   Summary of New York Control Area Generation Facilities' Energy, by Fuel Type, as of January 1, 2004, 101
D-2-9   Summary of New York Control Area Generation Facilities' Winter Capacity, by Fuel Type, as of January 1, 2004, 102
D-2-10   Summary of New York Control Area Generation Facilities' Summer Capacity, by Fuel Type, as of January 1, 2004, 103
D-3-1   Natural Gas Consumption for Electricity in Zones H Through K, 2003, 104
D-3-2   Natural Gas Consumption for Electricity in Zones H Through K, 2004, 104
D-3-3   Estimated Natural Gas (NG) Consumption of a 2,000 MW Combined-Cycle Unit with a 95 Percent Capacity Factor, 104
D-5-1   Electricity Cost from Coal with Emissions Controls, 108
D-6-1   Estimate of Potential Impact of Renewable Generation Technologies on Indian Point Service Area, 111
D-6-2   Quantitative Estimates of Wind Potential in Indian Point Zones, 113

D-6-3	Biomass Potential Applicable to Indian Point, 115
D-7-1	Estimated Distributed Photovoltaics in the Indian Point Service Area in the Accelerated Deployment Scenario, 118
D-7-2	Current and Projected Distributed PV Cost (2005 dollars), 120
D-7-3	Current PV Related Policies in New York State, 121
D-7-4	Accelerated PV Deployment Scenario for New York (2005 dollars), 123
E-1	Locational ICAP Requirements and Installed Capacity for NYCA in 2005-2006, 130
E-2	The Capacity Factors in 2003 of Major Generating Units in New York City and Long Island, 135
E-3	New Generating Units Proposed for the NYCA in 2004, 141
E-4	New Generating Units Proposed for the NYCA in 2005, 142
F-2-1	NYISO Initial Base Case Capacity Details Adopted for the MARS Analysis, 150
F-2-2	Electricity Generation Load and Capacity Representing NYISO Initial Base Case, 151
F-2-3	NYISO Initial Base Case—Qualifying Additions to Capacity (MW), 153
F-2-4	Committee's Screening Study—Early Shutdown with Assumed Compensation from Planned NYCA Projects and Added Energy-Efficiency and Demand-Side-Management Measures (MW), 154
F-2-5	Committee's Screening Study—End-of-License Shutdown with Assumed Compensation from Planned NYCA Projects and Added Energy-Efficiency and Demand-Side-Management Measures (MW), 155
F-2-6	NYISO Initial Base Case with Alternate New England Transmission Constraints—Projected NYCA Reliability Loss-of-Load Expectation (LOLE) and Reserve Margin, 155
F-2-7	Committee's Screening Study: Impact on Reliability and Reserve Margins of Shutting Down Indian Point Without Adding Compensatory Resources: Comparison of the NYISO Initial Base Case with Early-Shutdown and End-of-Current-License Shutdown Cases, 156
F-2-8	Committee's Screening Study: Impact on Reliability and Reserve Margins of Shutting Down Indian Point and Adding Compensatory Resources from Announced Projects, Beyond NYISO Initial Base Case (Table F-2-3): Comparison of Early Shutdown and End-of-Current-License Shutdown, 157
F-2-9	Reference Case: Illustrative Additional Resources Beyond the NYISO Initial Base Case to Meet Load Growth and Scheduled Retirements and Ensure Reliability Criteria Are Met, and Including Reliability Results If Indian Point Is Closed Without Further Compensation, 158
F-2-10	Early Shutdown of Indian Point with Compensatory Resources, Case b2, 159
F-2-11	End-of-Current-License Shutdown of Indian Point with Compensatory Resources, Case c2, 160
F-2-12	Early Shutdown of Indian Point with High-Voltage Direct Current (HVDC) Cable, Case b3, 161
F-2-13	End-of-Current-License Shutdown of Indian Point with Compensatory Resources Including 1,000 MW HVDC Transmission Lines, Case c3, 162
F-2-14	Early Shutdown of Indian Point with Higher Efficiency and Demand-Side Management, Case b4, 163

F-2-15 End-of-Current-License Shutdown of Indian Point with Higher Efficiency and Demand-Side Management, Case c4, 164

F-2-16 Early Shutdown Without Compensatory Resources Beyond the Reference Case—Impact on NYCA Reliability (Loss-of-Load Expectation) and Reserve Margin, Case b1, 165

F-2-17 End-of-Current-License Shutdown Without Compensatory Resources Beyond the Reference Case—Impact on NYCA Reliability (Loss-of-Load Expectation) and Reserve Margin, Case c1, 165

F-2-18 Committee's Reference Case—Impact on NYCA Reliability (Loss-of-Load) Expectation and Reserve Margin, 165

F-2-19 Early Shutdown with Additional Compensatory Resources—Impact on NYCA Reliability and Reserve Margin, Case b2, 165

F-2-20 End-of-Current-License Shutdown with Additional Compensating Resources—Impact on NYCA Reliability and Reserve Margin, Case c2, 166

F-2-21 Additional Compensatory Resources, Including 1,000 MW North-South HVDC Transmission Line—Impact on NYCA Reliability and Reserve Margin, Cases b3 and c3, 166

F-2-22 Additional Compensatory Resources, Including Higher Energy Efficiency and Demand-Side-Management Penetration—Impact on NYCA Reliability and Reserve Margin, Cases b4 and c4, 166

F-2-23 Projected Impact on the Annual Variable Cost of Operation for the Northeast Region, NYCA, and Zones H Through K: All Scenarios, 2008-2015, Including Percentage Change from Benchmark of 2008 NYISO Initial Base Case, 167

G-1-1 Economic Potential: Annual Savings (in megawatt-hours) for Top Eight Residential Energy-Efficiency Measures—Zones J and K, 2007, 2012, and 2022, 170

G-1-2 Economic Potential: Annual Savings (in megawatt-hours) for Top Ten Commercial Energy-Efficiency Measures—Existing Construction End Use in Zones J and K, 2007-2022, 170

G-4-1 Current and Projected Distributed PV Cost, 177

G-4-2 Accelerated PV Deployment Scenario for the New York City Area, 177

FIGURES

S-1 New York Control Area load zones, 4

1-1 The New York Control Area high-voltage transmission network, 10
1-2 Average daily load and peak hour load in New York City, 11
1-3 New York Control Area load zones, 12
1-4 Generating capacity in the NYCA, by fuel type, 2005, 13
1-5 Capability of generating plants by NYCA zone and generator type, 13

2-1 Past and projected trends in real residential electricity price in New York state relative to 1980, 20
2-2 Effects of demand-reduction programs on daily power demand, 21
2-3 Global photovoltaic market evolution, by market segment, 1985 to 2004, 23
2-4 Phased-in programmable potential for expanded demand-side options in the Indian Point service territory (in megawatts of peak reduction), 30

4-1 Projections made by NYISO in 2004 and 2005: summer reserve margin for generating capacity in the New York Control Area, 49

5-1 NYISO reliability projections, 61
5-2 Approximate additional resources needed, 63
5-3 Impact on NYCA reliability loss of load (LOLE) of shutting down Indian Point without additional resources beyond the reference case, 63
5-4 Capacity assumed to meet load growth and compensate for retiring Indian Point, 66
5-5 Loss-of-load expectation after compensation, 66
5-6 Projected reserve margin for End-of-License (EOL) Shutdown of Indian Point with Compensation (Case c2), 67

D-4-1 Proposed Northeast pipeline projects, 105
D-5-1 Emissions control options for coal-fired generation, 106
D-5-2 Past and projected U.S. emissions from fossil power generation, 1965 to 2030, 107
D-5-3 Types of power plants, 108
D-7-1 Global PV market evolution by market segment, 1985 to 2004, 119
D-7-2 An accelerated PV market development path for New York (all estimates are 2005 dollars), 122

E-1 North American additions in historical perspective, 126
E-2 Locational installed capacity requirements for Long Island and New York City for 2005-2006, 130
E-3 Average total cost of production (in dollars per megawatt-hour generated) for a representative peaking unit, 132
E-4 Daily zonal spot prices ($/MWh), January 2000 to July 2005, for New York City in the balancing (real-time) market at 2:00 p.m. on the first day of each month shown, 133
E-5 Average price-duration curves in the balancing market for May-April in New York City (in dollars per megawatt-hour) for 2000-2001, 2002-2003, and 2004-2005, 134
E-6 Projections made in 2004 and 2005 of the summer reserve margin for generating capacity in the New York Control Area, 140

G-4-1 Accelerated PV market development path for the New York City area, 178

BOXES

1-1 Keeping Competitive Markets Operating, 9

4-1 The Cost of Replacing Indian Point: In Theory, 45
4-2 The Cost of Replacing Indian Point: In Practice, 46

5-1 Reliability Criteria, 60
5-2 Multi-Area Reliability Simulation (MARS) Model, 62
5-3 Multi-Area Production Simulation (MAPS) Software Model, 68

Abstract

This report presents the work of the Committee on Alternatives to Indian Point for Meeting Energy Needs. It reviews various options that are available for replacing the 2,000 megawatts of energy produced by the two nuclear reactors at Indian Point and assesses some of the requirements and impacts of installing the options in an appropriate time frame.

The Indian Point Energy Center is a key part of the electric power system that serves New York City and densely populated surrounding areas. Maintaining reliability of electric supply in the area is essential.

Even with Indian Point operating, new capacity will be needed to meet expected growth in the region and to replace other generating plant retirements. Replacing the two operating Indian Point generation units would add to the complexity of the task. Options are constrained by various technological, regulatory, financial, and infrastructure factors that must be considered in planning for a reliable electric energy supply for southeastern New York State.

Based on all of the information available to it, the committee identified no insurmountable technical barriers to the replacement of Indian Point's capacity, energy, and ancillary services. However, significant financial, institutional, regulatory, and political barriers also would have to be overcome to avoid threatening reliability. As this report discusses, many replacement options exist, and if a decision were definitely made to close all or some part of Indian Point by a date certain, the committee anticipates that a technically feasible replacement strategy for Indian Point could be achievable. A replacement strategy would most likely consist of a portfolio of the approaches discussed in this report, including investments in energy efficiency, transmission, and new generation.

While the committee is optimistic that technical solutions do exist for the replacement of Indian Point, it is considerably less confident that the necessary political, regulatory, financial, and institutional mechanisms are in place to facilitate the timely implementation of these replacement options. The importance of this issue cannot be overstated in developing options for maintaining a reliable electric energy supply for the New York City metropolitan area. The report discusses in greater detail various aspects of this challenge and includes specific conclusions and findings.

Summary and Findings

This report presents the work of the Committee on Alternatives to Indian Point for Meeting Energy Needs. For over a year, the committee reviewed a wide range of potential options and assessed the feasibility of implementing these options on a scale and a timetable sufficient to replace the capacity, energy, and essential ancillary services now provided by the two operating nuclear reactors at Indian Point.

The committee recognizes the magnitude and the complexity of the issue that it was asked to study. Indian Point Units 2 and 3 provide about 2,000 megawatts (MW) of baseload generating capacity in one of the most densely populated areas in the nation. Its output represents 11 percent of the total generating capacity in southeastern New York (i.e., Long Island, New York City, and Westchester County) and 23 percent of the electric energy delivered in this region.

Based on all of the information available to it, the committee has identified no technical obstacles that it believes present insurmountable barriers to the replacement of Indian Point's capacity, energy, and ancillary services. As this report discusses, a wide and varied range of replacement options exists, and if a decision were definitely made to close all or some part of Indian Point by a date certain, the committee anticipates that a technically feasible replacement strategy for Indian Point would be achievable. Replacements for Indian Point would be in addition to generating and transmission capacity needed for expected growth in the region and because of other plant retirements.

The report does not propose a "single solution" to the replacement of Indian Point. That was neither the committee's directive nor its mission. Indeed, from the committee's analysis, no "right" or clearly preferable supply alternative to Indian Point emerged. A replacement strategy for Indian Point would most likely consist of a portfolio of the approaches discussed in this report, including investments in energy efficiency, transmission, and new generation.

While the committee is optimistic that technical solutions do exist for the replacement of Indian Point, it is considerably less confident that the necessary political, regulatory, financial, and institutional mechanisms are in place to facilitate the timely implementation of these replacement options. The importance of addressing the nontechnical barriers cannot be overstated in developing options for maintaining a reliable electric energy supply for southeastern New York State. The report discusses in greater detail various aspects of this challenge and includes specific conclusions and findings.

Reliability is a key consideration, especially during peak demand. Adequate generating and transmission capacity exists to replace Indian Point during nonpeak hours, although costs might be significantly higher because Indian Point is the low-cost baseload unit. Reliability of power supply depends on several factors, including fuel availability, generation reserve, peaking load, and the growth in electric demand, both locally and regionally. An element of a reliable electricity supply also involves the stability of the transmission-distribution system. In general, the electric system in the Northeast is carefully balanced to account for the location and operation of baseload generating plants, as well as peaking units. In southeastern New York, the reliability criteria also impose specific locational resource requirements, reflective primarily of New York City and Long Island's situation as very large demand centers at the end of the transmission grid. For these reasons, the committee's analysis has focused on replacement strategies, that is, on electric energy supply and demand options, primarily in southeastern New York (Zones H, I, J, and K; see Figure S-1).

Adding to the complexity of choice is the issue of cost to customers and taxpayers, which could include the costs of both closing Indian Point and providing replacement resources. For example, if the plant's life were shortened, compensation might be owed to the owner. Costs of maintaining site security would be required to keep the spent nuclear fuel secured. There is considerable uncertainty over how the cost of replacement resources, higher fuel prices, and air quality

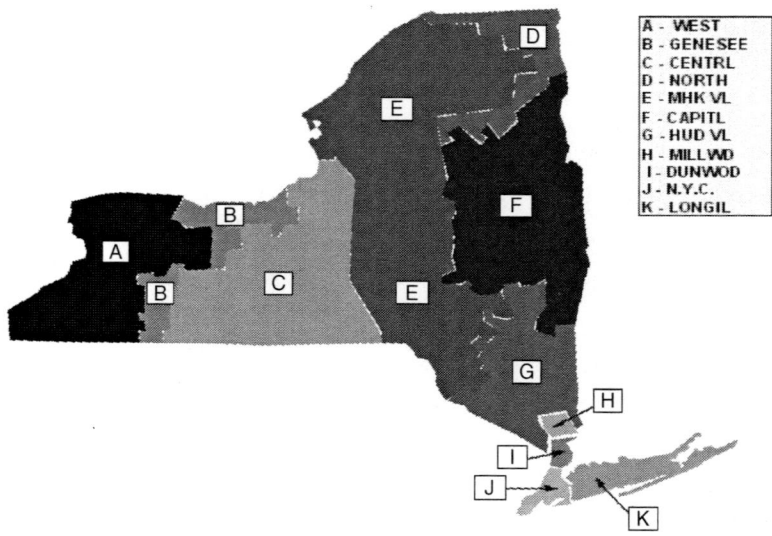

FIGURE S-1 New York Control Area load zones. SOURCE: New York Independent System Operator.

offsets would be addressed in a deregulated wholesale electric market in which price is no longer based on the cost of production but rather on an open competitive bidding process under which all bidders get the same price as the last successful marginal winning bid. Also of concern are potential indirect costs to the community at large and state and local governments, including any loss of tax base from the plant, labor dislocation, or loss of income from reduced plant operations that might be associated with the closure of the Indian Point facility.

Indian Point sits on the banks of the Hudson River whose protection has been a focal point of the American environmental law movement, so it is no surprise that a complex web of federal and state environmental regulations must also be considered in evaluating replacement resources for Indian Point. These include air quality, water quality, and thermal discharge requirements; regulations regarding toxic releases; and regional and perhaps eventual federal initiatives to reduce greenhouse gas emissions. New power plants can be permitted only under the most stringent environmental review processes, and such projects are also subject to local zoning and land use controls.

CONCLUSIONS AND FINDINGS

The issues associated with the potential shutdown of Indian Point's two operating nuclear units are complex and interrelated. These issues impact the total energy system for New York State, the Northeast region, and beyond. Any analysis of the consequences and potential alternatives to the closure of Indian Point units cannot occur in a vacuum without reference to the context of other events unfolding in the state.

In analyzing replacement options for Indian Point, the committee examined the broader profile of New York State's electric power system to identify what, if any, other existing resources might be available to replace some portion of the energy and capacity now provided by Indian Point. Most germane to its evaluation of replacement options for Indian Point, the committee learned that even with the Indian Point units operational, New York State will require system reinforcements, above those already under construction, as soon as 2008 in order to meet its projected demand for electricity and maintain system reliability in the Lower Hudson Valley and New York City area served by the Indian Point units. The state's need for additional electric power resources increases rapidly thereafter. Based on currently scheduled retirements and demand growth projections by the New York Independent System Operator (NYISO), 1,200 to 1,600 MW from new projects that are not yet under construction could be needed by 2010, and a total of 2,300 to 3,300 MW by 2015. Closing Indian Point would increase by 2,000 MW New York's need for additional electric resources, which could be in the form of new generating capacity, transmission lines, improved energy efficiency, and demand-side management.

This need for new resources is occurring at a time when it is problematic whether the existing legal, regulatory, and financial mechanisms provide sufficient incentive to build new resources in New York. The committee estimates that the generating capacity currently under construction will be insufficient to meet projected peak demand in 2009, given currently announced retirements. With the expiration in 2003 of its siting statute, Public Service Law Article X, New York State has no law designed to facilitate an integrated environmental review and siting of new power plants. NYISO has just completed its first Comprehensive Reliability Planning Process, and as this report explains in detail, it remains to be seen whether NYISO's new market and pricing rules will provide sufficient economic incentives to stimulate investment in new electric resources. Developers and financial markets will look for investment opportunities with the best combination of high payback and low risk, whether they are in New York or not. If price signals in New York are low, the markets will wait until they rise. Given the time that it takes to obtain a suitable site, navigate the regulatory issues and obtain permits, and then construct a power plant, new generating capacity may not be available until reserves are dangerously low. Forestalling a crisis may require extraordinary efforts on the part of policy makers and regulators.

The committee examined two time frames for the possible closure of Indian Point: (1) when the current operating licenses expire for the two reactors in 2013 and 2015; and (2) an accelerated schedule of 2008 and 2010. The general conclusions that the committee reached concern the overall ability to replace the capacity and energy required if the Indian Point units were shut down in either of the two time frames. The committee also reached agreement on eight specific findings associated with generation, transmission, and demand-side options; reliability; physical and political infrastructure; the environment; and cost considerations if an early shutdown of Indian Point is effected. The committee emphasizes that the inability to successfully meet any of the requirements set forth in its eight findings would place the general conclusions in jeopardy.

General Conclusion (2013-2015)

The committee concludes that with sufficient time, planning, authority, and investment incentives, options are possible for replacing Indian Point. The Indian Point units could be retired at the end of their current operating licenses (2013 and 2015) without causing a major disruption of power capacity in southeastern New York if sufficient resources were added by 2015 to cover anticipated system retirements and the expected growth in demand, as well as the shutdown of Indian Point. To achieve this goal, the committee estimates that an additional 5,000 to 5,500 MW, or roughly 500 MW per year, in new resources (a combination of generation, transmission, and demand-side actions) would need to be added by 2015.[1] The 3,300 MW in new resources that are estimated to be required even if Indian Point continues to operate is less than 10 percent of New York's current capacity, and it should be achievable over the next 9 years. The additional 2,000 MW of new resources required if Indian Point is closed should also be achievable if the conditions discussed below are met.

General Conclusion (2008-2010)

The committee concludes that an earlier shutdown of the Indian Point units would be much more difficult to accomplish. In 2008, when Unit 2 (1,000 MW) would be closed, New York will have very little if any excess capacity. To replace it, the committee estimates the need for an additional 700 MW in generating capacity, assuming that demand-side programs could reduce peak demand by several hundred megawatts. By 2010, with the closure of the second unit (1,000 MW), an additional 1,300 to 1,400 MW in replacement generating capacity would be needed, assuming that demand-side measures would continue to increase, totaling 650 MW in peak-demand reductions. That is in addition to the 1,200 to 1,600 MW that will be needed even with Indian Point operating. In the committee's view, this extraordinary challenge could only be met with the firm commitment of a variety of New York government leaders and tight cooperation among many agencies. Such collaboration may be unprecedented, so the difficulty of achieving it should not be underestimated. The impacts discussed for the 2013-2015 scenario would be magnified, with potentially even greater added costs. If new generating capacity is not constructed in a timely manner, system reliability would be threatened. Not only could reserve margins drop below standards, but existing generating units would likely show lower reliability as they are run beyond their normal operation schedule.

Finding 1: Governmental Mechanisms and Regulatory Policy

The committee recognizes that maintaining a reliable supply of electricity for New York City and southeastern New York State is a primary objective for public policy and essential to the region's health and economic well-being. However, the committee finds that current governmental mechanisms and regulatory policy may limit New York State's ability to address in a timely and effective manner the capacity, energy, and ancillary consequences of closing Indian Point. The committee finds that in order to provide alternatives to Indian Point Units 2 and 3, a more considered long-range strategy is likely to be necessary. This strategy would be based on a detailed assessment of the current market structure and might well require significant changes in New York's current laws and regulatory policies, such as reauthorization of the state's Article X power plant siting process and reestablishment of the State Energy Planning Board and

---

[1] All projections in this report should be understood to be approximate at best. Not only are estimates of load growth uncertain, but assumptions of where new generating and transmission capacity will be added, constraints on system operations, and the analytical methodology that is used would all affect the estimates of reliability and the calculated need for new capacity.

the state energy planning process, in order to ensure the continued reliability of the state's electric system.

Finding 2: Market and Financial Uncertainties

The committee notes that even with the continued operation of the Indian Point units, New York State already faces challenges in satisfying the projected growth in its electric demand and in maintaining system reliability. While conceptual planning to address these needs is under way through the New York Independent System Operator and other entities, the response of electric power developers, suppliers, and distributors is uncertain, given the current state of evolution of New York's market. Indian Point represents a significant asset, both in terms of capacity and energy, especially for electric customers in southeastern New York, and if Indian Point is retired, replacement of its 2,000 MW capacity will place a substantial additional burden on the state's electric supply system.

Finding 3: Transmission Options

The committee finds that improvements in transmission capability could significantly relieve congestion in the New York system and facilitate the delivery of power from existing and potential electric generation resources to the New York City area. Such improvements should include modifications to the state's existing transmission system and the possible installation of new direct current transmission. A West-to-East line (550 MW) has been proposed across the Hudson River, and a new North-to-South transmission line (up to 1,000 MW) for better access to upstate and Canadian electric resources is under investigation. These lines could supply useful capacity in the 2010 and 2015 time period, respectively, if a variety of institutional and financial issues can be resolved. The committee notes that increasing the importation of power into southeastern New York would also increase the need to install additional reactive power equipment to maintain system voltage within the region, but this problem is relatively easy to solve.

Finding 4: Demand-Side Options

The committee finds that substantial cost-effective opportunities exist for investment in demand-side technologies that could reduce demand for electricity in southeastern New York. These could include a phase-in of programmable energy efficiency and demand-response programs, along with additions of distributed generation and combined heat and power units. These could provide reductions of more than 1,100 MW from projected peak demand by 2010 and 1,700 MW by 2015. The committee notes that these offsets are ambitious and would be in addition to the current effective programs with which the New York State Energy Research and Development Authority, the New York Power Authority, Consolidated Edison, and the Long Island Power Authority are already managing demand growth. The committee finds that these offsets are achievable, but only if well-designed programs are implemented promptly and additional resources are provided to overcome many obstacles.

Finding 5: Supply-Side Options

The committee finds that even with substantial additional investment in new transmission facilities and aggressive demand-side programs, additional generating facilities, above those already planned, would be required to compensate for the shutdown of the Indian Point units to maintain system reliability. While coal may be a reasonable generating alternative for the 2013-2015 time frame, new near-term generating solutions are most likely to be a mix of simple-cycle gas turbines and combined-cycle natural gas units. The use of the former would provide a short-term solution, but in the longer term, such units would probably be relegated to peaking usage. Owing to the nature of the New York City metropolitan region, renewable energy technologies are unlikely to contribute significant resources by 2015, with the possible exceptions of offshore wind power and distributed photovoltaics.

Finding 6: Alternative Fuel Availability and Security

The committee finds that the availability and price of natural gas would be major considerations, and perhaps constraints, in planning for new generating capacity to replace power from the Indian Point units. A large share of the 2,000 MW from Indian Point would likely be replaced with natural-gas-fired generating plants, and that is over and above the several thousand megawatts of new gas-fired capacity that will be needed to meet the growing demand for energy in southeastern New York State. This increase in New York's dependence on natural gas for power production will stress supplies of natural gas. In addition, increased dependence on natural gas will reduce diversity of fuel supply for the New York electric system, also a serious concern.

Finding 7: Cost Considerations

Cost is a key consideration in evaluating any scenario for the early retirement of the Indian Point units. Three main categories must be taken into account: (1) any compensation that might be due Entergy Nuclear for the early retirement of the Indian Point units; (2) replacement costs, including new generation and transmission, demand-side programs, increased demand for pollution offsets, and the increased price of fuel, particularly natural gas for power production; and (3) the financial impact to Westchester County, the Town of Buchanan where Indian Point is located, and surrounding communities from the loss of Indian Point tax revenues and the labor-commercial base. The committee found that it is difficult to make specific cost estimates for these items. Ultimately, the price that consumers pay for electricity in southeastern New York will reflect some of these costs. However, given the current market structure for the sale of electric power in New York, under which wholesale prices are set on a subregional zonal basis that reflects competitive bidding behavior, the committee could not satisfactorily determine the increase in the cost of electricity to consumers that might result from the closure of Indian Point. Some costs could be offset by demand-management practices, but new generation, and perhaps new transmission, will likely increase wholesale electric costs, especially in the New York City metropolitan area, depending on competitive bidding in the open wholesale market.

Finding 8: An Integrated Approach Is Needed

The committee emphasizes that its findings must be considered as an integrated whole. Replacements for the energy, baseload capacity, and ancillary services currently provided by the Indian Point units will not happen just because they should. The construction and operation of new electric generating facilities, natural gas pipelines, liquefied natural gas facilities, or electric transmission lines will each inevitably encounter hurdles that will have to be overcome if that project is to become a reality. Each facility needs a site, financing, permits, delivery contracts, and infrastructure agreements, and has facility-specific requirements. This is also true for any demand-side programs, which have their own timing, financial, marketing, and implementation challenges to be worked out in order to achieve sufficient participation by the general public.

1

Introduction

This report presents the work of the National Research Council's (NRC's) Committee on Alternatives to Indian Point for Meeting Energy Needs. It reviews the options that are available and assesses the feasibility of installing them on a scale sufficient to replace the 2,000 megawatts (MW) of electricity from the Indian Point Energy Center.

This chapter presents background information necessary to understand how replacements would be implemented. It also reviews how the committee conducted the analysis.

BACKGROUND

Electricity Supply and Demand

Electricity generally cannot be stored and must be generated at virtually the same instant as it is used, which requires continuous control of the system.[1] New York State has an integrated bulk power system, the New York Control Area (NYCA). Formerly, the New York Power Pool had coordinated the activities of the utility participants on the transmission system. As competition was introduced into the New York electric system, utilities were required to divest their generating assets.[2] The New York Public Service Commission and the Federal Energy Regulatory Commission also required a more independent electric system operator. The New York Independent System Operator (NYISO) was created to operate the high-voltage transmission system and to provide a match of load requirements to generation sources in a manner that (1) ensures the reliability of the state's power system; (2) facilitates open, fair, and effective competitive markets; (3) improves regional cooperation for operations and planning; and (4) ensures nondiscriminatory access to the electric system.

NYISO uses the locational-based marginal pricing (LBMP) system to accomplish its objectives. LBMP also provides price signals to providers of new generation and transmission. Thus, NYISO has assumed the power-dispatching role that integrated utilities used to carry out within their own jurisdictions, but on a statewide level. NYISO uses auctions to select the lowest-cost suppliers consistent with transmission constraints, among other functions. Box 1-1 lists many of the market products that NYISO must monitor. Further details are provided in Chapter 4. Competitive markets are still evolving, and it is not yet clear exactly how to ensure both reliability and low costs.[3]

NYISO also plans for future growth and makes recommendations for additional capacity, although it does not pick specific sites or technologies. Additional capacity is mainly built by developers, or merchant generators, which could have contracts for the power from a load serving entity (LSE) or which expect to be able to compete profitably in the auction. Under some conditions, the New York Power Authority (NYPA) can build new capacity. NYISO has issued a request for proposals to deal with concerns over potential capacity shortfalls, but that process has just begun.

---

[1] Pumped storage facilities, currently the only practical form of large-scale power storage, use low-cost off-peak power to pump water uphill to a reservoir. The flow is reversed during peak hours when the power that can be regenerated is much more valuable. However, few sites are appropriate for pumped storage. Consolidated Edison attempted to build pumped storage on Storm King Mountain up the Hudson River near West Point, but the project was stopped for environmental reasons. Other storage technologies, including batteries, compressed air energy storage, and superconducting magnets, are still under development to reduce costs.

[2] Competition was introduced in part to avoid cost increases, such as had occurred in the 1970s and 1980s because of overbuilding. Those costs had largely been passed on to customers.

[3] Competitive markets, or "restructuring," encompass (1) allowing generation to be built by nonutilities; (2) breaking up vertically integrated utilities; (3) independently owned and operated transmission, with some degree of open access for all suppliers; (4) spot markets for electricity; (5) retail choice for some customers in some states (including New York); and (6) a substantial shift in regulatory jurisdiction from the states to FERC. They may also include competitive bidding for power supply and the inclusion of energy efficiency in competitive power procurement processes.

INTRODUCTION

> **BOX 1-1**
> **Keeping Competitive Markets Operating**
>
> New York's large and varied power system requires a very complex set of functions for smooth and efficient operation. NYISO conducts energy market auctions in two phases: (1) the Day Ahead Market establishes forward contracts for each hour of the coming day; (2) the Real Time Market is conducted when the load actually occurs to precisely match supply with demand. Most energy transactions in NYISO are conducted in the Day Ahead Markets. NYISO adds up the bids starting with the lowest cost for each time interval until it has sufficient power to meet projected demand. All bidders then receive the price set by the highest accepted bidder.
>
> Other important functions include the Installed Capacity (ICAP) Market, which is designed to ensure that load serving entities (LSEs, such as ConEd) have sufficient capacity available to serve their customers. The following are among the NYISO market products, as described in detail on the NYISO website (www.nyiso.com):
>
> **Energy Markets**
> Day-ahead locational-based marginal pricing (LBMP) energy
> Real-time LBMP energy
>
> **Ancillary Services**
> Regulation service (frequency control)
> Black start capability
> Voltage support service (reactive power)
>
> **Installed Capacity (ICAP)**
>
> **Transmission Congestion Contracts**
>
> **Demand Response Programs**
> Emergency Demand Response Program
> Special Case Resources (SCR)
> Day Ahead Demand Response Program
>
> SOURCE: www.nyiso.com; accessed March 29, 2006.

Reliability standards are set by the New York State Reliability Council (NYSRC) in conjunction with the Northeast Power Coordinating Council (NPCC), which operates under the North American Electric Reliability Council (NERC). NPCC standards also apply to New England and eastern Canada, while NYSRC standards are tailored to New York's particular situation (e.g., requirements for generating capacity in New York City and Long Island). NYSRC also sets the amount of installed generating capacity (ICAP) needed to meet the required reserve margin generating capacity at peak electrical load. Reserve margin criteria are set yearly for 1 year ahead (18 percent for 2006-2007) by NYSRC, which also specifies other allowable resources (e.g., specific loads that can be shut off on NYISO's order are equivalent to generating capacity for meeting peak demand) to be included in the reserve margin and correspondingly to be used in calculating the reliability. Finally, the Energy Policy Act of 2005 provides that the Federal Energy Regulatory Commission (FERC) will certify a single organization (expected to be NERC) that will propose and enforce mandatory "reliability standards for the bulk-power system in the United States," subject to FERC approval.

A complicated network of high-voltage transmission lines is required to deliver the bulk power to load centers, which may be hundreds of miles from the generating stations.[4] The bulk power system must be controlled very precisely to keep voltage and frequency within tight bounds and to operate reliably despite the occasional component failure. It also is important to keep the cost of electricity as low as possible, in part by operating the lowest-cost plants as much as possible.

The NYCA has about 38,000 MW of installed capacity within New York State and 4,000 miles of high-voltage transmission lines. Power also can be traded with interconnected control areas in New England, the Mid-Atlantic region, and Canada. The NYCA high-voltage transmission system, including major substations, is shown in Figure 1-1.

Power demand fluctuates both during the day and over the year, as shown in Figure 1-2, so a variety of generating plants must be available to follow the load, including:

- *Baseload plants, to meet the steady part of the load.* Baseload facilities (such as the Indian Point units) produce power inexpensively. They typically operate all day and most of the year. They are generally nuclear or coal-fired steam generators. The Indian Point units are an important generating resource in the NYCA owing to their low cost and their location near the load centers in New York City and Westchester County.
- *Peaking plants for periods of high demand.* Combustion turbines, for example, are often deployed in simple cycle, and are used during periods of peak demand, because they can be quickly turned on or off. The operational flexibility of such "peaking" generators, however, is counterbalanced by their low thermal efficiencies, which makes them expensive to operate.
- *Intermediate units, which also follow demand but are used more than peaking plants.* An intermediate generator might use a combustion turbine in combination with a steam turbine to provide a wide range of operating flexibility. Combined-cycle facilities are typically fueled with natural gas and often have the capability of burning oil as an alternative fuel supply when supplies of natural gas are curtailed because of high demand, usually during the winter. Modern

---

[4]Low-voltage distribution lines, which are not part of the bulk power system, carry the power to the end-use customer. Most outages that consumers experience are due to failures in the distribution system (e.g., trees falling on overhead lines), but these usually are repaired quickly and are not part of this study.

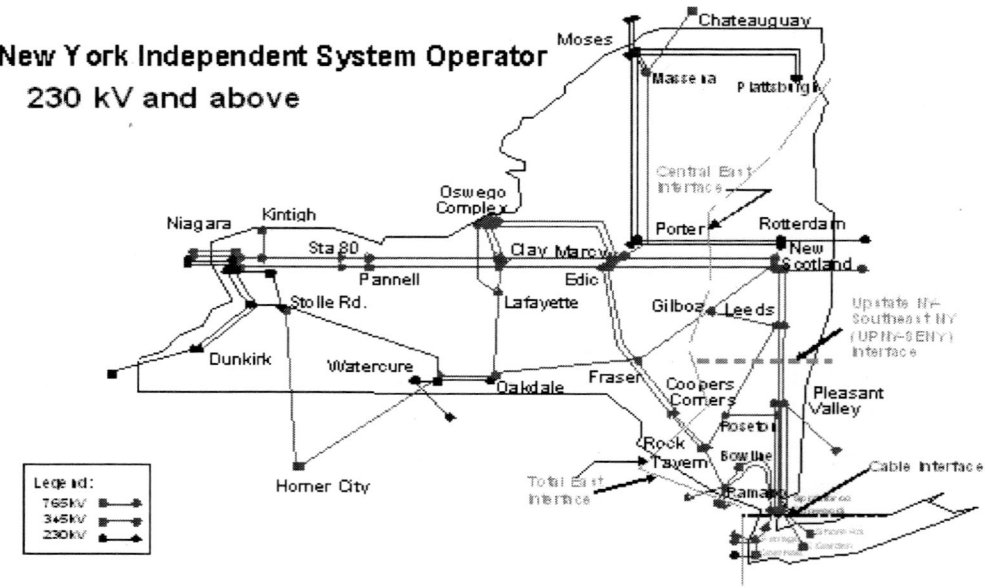

FIGURE 1-1 The New York Control Area high-voltage transmission network. SOURCE: New York Independent System Operator.

gas-fired combined-cycle plants[5] are much more efficient than older or simple-cycle gas turbines.

NYISO has divided the NYCA into 11 zones, shown in Figure 1-3, to assist in pricing and monitoring load flows on the transmission system. The key zones for this report are these:

- H, which includes the northern portion of Westchester County, where Indian Point is located;
- I, the rest of Westchester County;
- J, New York City; and
- K, Long Island outside of New York City.

In accordance with NYSRC standards, NYISO's goal is for the bulk power system to have sufficient capacity that outages will be less than 1 day in 10 years. This loss-of-load expectation (LOLE) is determined by using statistical descriptions of the historical availability of each generator and Monte Carlo calculation techniques to compute the expected number of days in a 10-year period when the load could not be supplied. The LOLE is used in determining how much additional generation a given area will require for expected load growth and is likely to continue to be used if Indian Point is closed.

In addition to sufficient capacity, diversity of fuels provides another element of system reliability. Excessive dependence on one fuel source threatens system reliability if that fuel supply encounters shortages. Figure 1-4 displays the varied contributions of different fuels to the installed capacity (in megawatts) of the NYCA. Natural gas and oil represent 60 percent of the installed capacity, and coal, nuclear, and hydroelectric power account for 39 percent. New York's new Renewable Portfolio Standard should improve fuel diversity. This standard requires 25 percent of electricity to be generated from renewable sources by 2013, compared with 19.5 percent now (mainly hydroelectricity, most notably from Niagara Falls).[6]

The electrical output (actual kilowatt-hours) generated by each fuel is not proportional to the generating capacity that uses that fuel. Gas and oil fuel about 38 percent of the total. Coal, nuclear, and hydro power represents most (61 percent) of the power generated in 2004.

Generator owners in the NYCA operate a diverse mix of generation facilities. Figure 1-5 lists the power that can be generated in each NYCA zone, by technology, during the

---

[5]These plants combine a gas turbine (similar to a jet engine) with a steam turbine that uses the waste heat from the gas turbine as its energy source. The latest combined-cycle plants can be up to 60 percent efficient, almost twice as high as most coal or nuclear plants.

[6]Renewable resources include solar energy, wind, biofuels, and others. Renewables are appealing for a variety of reasons, especially environmental, but most forms have been expensive relative to fossil and nuclear energy. Some technologies (e.g., wind) are now proving to be competitive, and progress in research and development on others is encouraging, as discussed in Chapters 2 and 3. Hydroelectricity is a form of renewable energy, and New York State already receives an abundant supply from Niagara Falls and other sites, but it is questionable whether hydropower can be expanded significantly.

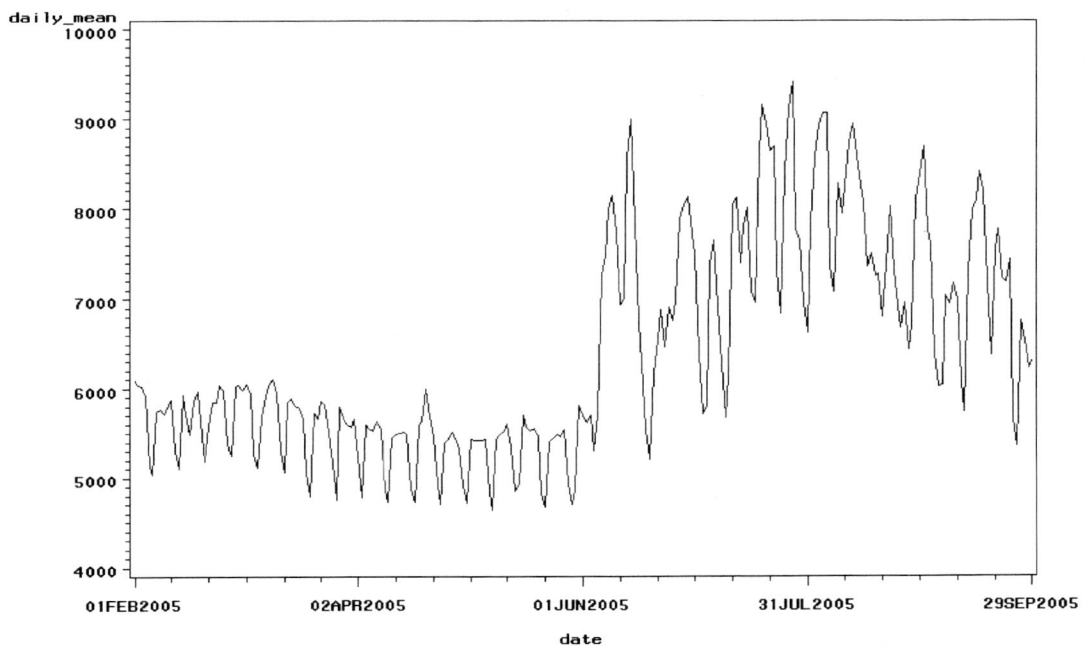

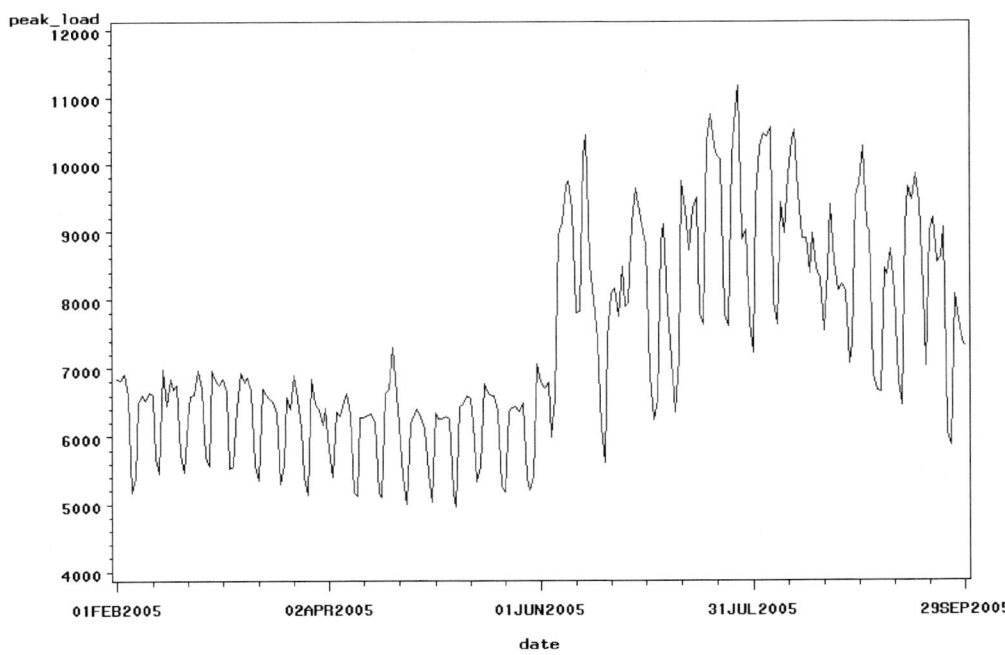

FIGURE 1-2 Average daily load (top) and peak hour load (bottom) in New York City. SOURCE: Personal communication with Timothy Mount, Cornell University, compiled from NYISO data, January 2006.

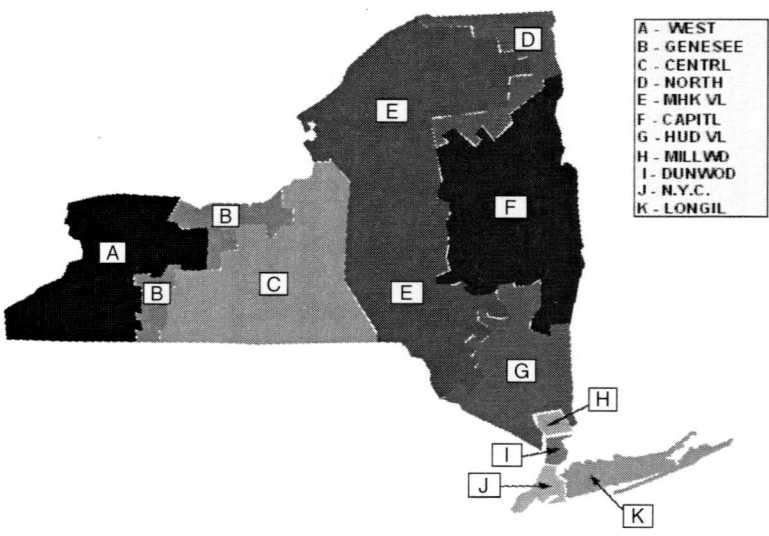

FIGURE 1-3 New York Control Area load zones. SOURCE: New York Independent System Operator.

summer-peak demand period.[7] The diversity of generator technologies in the NYCA in itself adds to the reliability of the electrical system. Reliability also is a function of the location of the generating facilities relative to the load centers that they serve. Indian Point Units 2 and 3 (total 1,970,700 kW) are listed in the column "Zone H" and row "Steam (PWR [for pressurized water reactor] Nuclear)." The two units represent 12.5 percent of the total summer capability in Zones H, I, J, and K (NYISO, 2005). Indian Point is virtually the only generating facility in Westchester County.[8]

Even with adequate capacity, an electric grid may fail because of instability. Several types of instability may occur, and they have different timescales and effects on customers. Voltage stability is most important in considering alternatives to Indian Point. The phenomenon of voltage collapse (in which voltage declines to unacceptable levels, as it did in Ohio in August 2003) is associated with insufficient reactive power.[9] The existing generators at Indian Point can supply a large amount of reactive power when it is needed. It will be necessary to verify that alternatives to Indian Point would include sufficient reactive power to maintain acceptable voltage levels under all predicted loads.

Peak demand generally occurs during hot summer afternoons when air conditioning loads are highest. Demand on July 26, 2005, was 32,075 MW, a record for the NYCA. Reliability is of greatest concern during hours of peak demand because at such times reserve capacity, both generation and transmission, is at its lowest. Any equipment failure then can threaten continued supply if reserve capacity is too low. NYSRC has a general requirement that NYCA capacity must exceed expected peak demand by 18 percent to allow for failures.[10] On July 26, the reserve margin was about 19 percent, indicating adequate reserve capacity for the state.

Regional distribution within the state, however, is more problematic. Upstate New York has some surplus capacity, but very little if any additional power can be delivered downstate because the transmission system is already congested during peak demand. Furthermore, electricity demand has been growing at over 2 percent per year in southern New York, so more capacity will be required in a few years to meet peak demand in that area. Chapter 2 includes an analysis of demand growth and the options for controlling it. Chapter 3 discusses the possibility of building new power plants upstate and transmission lines to bring the power south.

In addition to controlling bulk power flows, NYISO must monitor and control reactive power. Insofar as reactive power cannot be produced by operating generators, it must be supplied by specialized equipment.

Several other factors extremely important in planning for the future of the bulk power system noted here are discussed further in Chapter 3. A reliable supply of electricity depends on a reliable supply of fuel to power the generators. New York has a diverse supply of fuels: hydroelectric, nuclear,

---

[7] Many generating plants can produce more power in the winter than in the summer. Cooler air is denser, so combustion turbines can be fed more fuel. Steam turbines also exhaust to a lower temperature and thus lower back pressure, increasing their efficiency.

[8] Zone I has about 3 MW of hydroelectric power and municipal waste generation in addition to the 2,000 MW from Indian Point; see Appendix D-2 for details.

[9] Reactive power is a complex phenomenon in alternating current power. It is discussed further in Chapter 3 of this report.

[10] Reserve margin during off-peak hours is, of course, much higher. It is only high-demand hours that are of concern.

INTRODUCTION

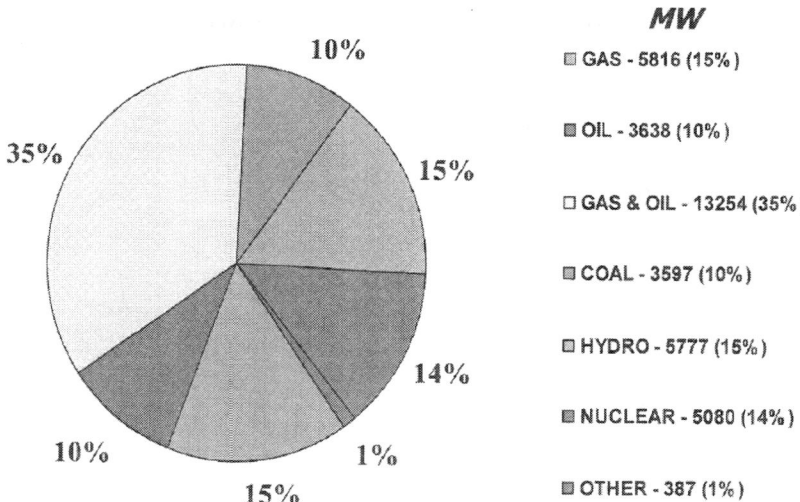

FIGURE 1-4 Generating capacity in the NYCA, by fuel type, 2005. SOURCE: New York Independent System Operator, *Power Trends*, April 2005.

coal, natural gas, and oil. Diversity is important because disruptions can occur in fuel deliveries. In recent years, most new generation has been fueled with natural gas, but new supplies of gas are expected to be limited and expensive unless new facilities for importing liquefied natural gas (LNG) are built. Natural gas is generally available during the summer, but it may be curtailed in the winter when demand is high for residential and commercial heating. Oil is frequently used as a backup for natural gas in the winter, but it is expensive, pollutes more, and raises national security issues.

Environmental factors may control what types of facilities can be built where. In particular, air pollution regulations can limit the use of coal, the nation's most abundant fossil fuel. New York has introduced new, lower standards

Capability By Zone and Type
As of April 1, 2005

Generator Type	ZONE A	ZONE B	ZONE C	ZONE D	ZONE E	ZONE F	ZONE G	ZONE H	ZONE I	ZONE J	ZONE K	TOTAL
	Summer Capability Period (KW)						Summer Capability Period (KW)					
Steam Turbine (Oil)	0	0	1649200	0	0	0	0	0	0	0	0	1649200
Steam Turbine (Oil & Gas)	0	0	0	0	0	0	2446100	0	0	4353100	2441700	9240900
Steam Turbine (Gas)	0	0	0	0	0	0	0	0	0	827900	238700	1066600
Steam Turbine (Coal)	1901700	238000	676500	0	52400	0	728300	0	0	0	0	3596900
Steam Turbine (Wood)	0	0	0	18100	20200	500	0	0	0	0	0	38800
Steam Turbine (Refuse)	37800	0	33116	0	0	11800	8200	52000	0	0	120800	263716
Steam (PWR Nuclear)	0	498800	0	0	0	0	0	1970700	0	0	0	2459500
Steam (BWR Nuclear)	0	0	2610000	0	0	0	0	0	0	0	0	2610000
Pumped Storage Hydro	240000	0	0	0	0	1048700	0	0	0	0	0	1288700
Internal Combustion	6612	2100	25212	1700	0	1718	13500	0	0	2000	65740	118582
Conventional Hydro	2396060	57018	122377	921514	459746	423669	105400	0	2200	0	0	4487984
Combined Cycle	462704	119100	1385100	320600	328600	1624300	0	0	0	1363300	239800	5843504
Jet Engine (Oil)	0	0	0	0	0	0	0	0	0	0	526800	526800
Jet Engine (Gas & Oil)	0	0	0	0	0	0	0	0	0	0	172600	172600
Combustion Turbine (Oil)	0	14000	0	0	0	0	15600	46500	0	784100	553900	1414100
Combustion Turbine (Oil & Gas)	0	0	0	0	0	0	104000	0	0	1185600	138400	1428000
Combustion Turbine (Gas)	38000	14000	85500	0	0	0	0	0	0	465400	681500	1284400
Wind	26	6700	30026	0	9865	20	10	0	0	0	0	46647
Other	0	0	0	0	0	0	0	0	680	0	0	680
Totals	5082902	949718	6617031	1261914	870811	3110707	3421110	2069200	2880	8981400	5179940	37547613

FIGURE 1-5 Capability of generating plants by NYCA zone and generator type. See Figure 1-3 for a map of NYCA zones. SOURCE: New York Independent System Operator, *Power Trends*, April 2005.

for emissions of sulfur dioxide and nitrogen oxides, which would require expensive emissions controls on coal plants. Carbon dioxide emissions are emerging as an issue. Concerns over global climate change are leading to restrictions on emissions of greenhouse gases, though not yet at the national level. New York is part of the recently adopted Regional Greenhouse Gas Initiative, which will begin to limit emissions of carbon dioxide in 2008.

The changing institutional structure of the electric power industry in New York will also play an important role in efforts to replace Indian Point, as described in detail in Chapter 4 and in Appendix E. Formerly, under the regulated approach, an integrated utility would determine its generating, transmission, and other needs, and build whatever was required. A reasonable return on its investments was largely guaranteed by the state's Public Service Commission. The introduction of competition in the industry has also introduced an element of uncertainty that affects the willingness of power companies to invest. The expiration of New York's siting legislation in 2003 represents another hurdle to building new facilities.

Finally, societal impacts play an important role in guiding decision making with respect to the bulk power system. These impacts can be seen in issues such as public opposition to new generating or transmission capacity. Employment issues can also be important for some facilities.

The Indian Point Energy Center: Description and Role

Three reactors have been built at the 239-acre Indian Point site. Unit 1 was an early, small reactor that has been shut down since 1974. It is still onsite though not operable, because demolition was deemed easier if carried out simultaneously with the later reactors.

Indian Point Unit 2 was built by Consolidated Edison (ConEd), the utility that supplies power to Westchester County and New York City. Operating since 1974, Unit 2 is licensed until September 28, 2013. Until recently it produced 970 MW but has now been upgraded to 1,078 MW.

Construction of Indian Point Unit 3 was started by ConEd, but financial difficulties forced the utility to sell it to NYPA before completion. It has operated at 980 MW since 1976 and is licensed until December 12, 2015. It has now been upgraded to 1,080 MW.

In 2001 and 2002, the units were sold to Entergy Corporation, an integrated energy company that owns and operates power plants. Both sales were accompanied by an agreement to purchase back the power generated by the plant for several years. These agreements are phasing out, and Entergy will soon be able to sell the power at a higher price, as most alternate fuels have risen considerably in cost over the past few years.

Entergy Nuclear operates 10 nuclear power plants, including the Indian Point Energy Center and the FitzPatrick plant in upstate New York. Since Entergy took over Indian Point, it has operated the plants extremely well. From 2003 to 2005, Unit 2 operated at a capacity factor of 96.6 percent and Unit 3 at 93.7 percent (NEI, 2006). The industry average is 89.6 percent. The two Indian Point reactors are among the lowest-cost generators in New York, and they operate whenever possible supplying base load power to the system. Together, they account for 5.3 percent of the total installed generating capacity in New York State, but they produce 10.1 percent of the electricity (Levitan and Associates, 2005).

Entergy can apply for license extensions for an additional 20 years of operation. The U.S. Nuclear Regulatory Commission would review the applications for confirmation that the reactors could be operated safely and in compliance with environmental regulations. The application process can take about 5 years, suggesting that Entergy would have to submit the applications for Units 2 and 3 in 2008 and 2010, respectively.

Both units feed power into the transmission network at the nearby Buchanan substation. The power is delivered to load centers, mainly in New York City.

Indian Point is the largest generating station close to the major load centers in New York City, Westchester County, and Long Island and south of congestion points in the NYCA transmission system that prevent more power from being sent south during periods of peak demand. Indian Point also produces the lowest-cost power in the area. Thus, Indian Point is a critical component of both the reliability and economics of power for the New York City area. In addition, it produces much of the reactive power needed for reliable operation of the system. Replacing Indian Point will call for careful analysis of the choices that are made.

Community Concerns

Community concerns about the Indian Point reactors have a long history (Wald, 1982), but prior to September 11, 2001, they had faded, with only a few people still expressing public concern that the dangerous amounts of radioactivity in the cores of the reactors might be released in an accident (Hu, 2002). Opinions were changed by the 2001 attacks on the World Trade Center (Purdy, 2003; Lombardi, 2002; Hu, 2002).

> Since the Sept. 11 terrorist attacks, growing anxiety over the safety of nuclear power plants has transformed Indian Point from a fringe issue that only antinuclear crusaders care about to a mainstream concern, and not just for Westchester suburbanites, but for New York City and New Jersey residents, who had, until now, barely registered the plant's existence 40 miles north of Midtown Manhattan. (Hu, 2002)

Scenarios leading to catastrophic releases were no longer easy to dismiss on the basis of fault-tree calculations and experience underlying previous assurances of safety, although the Nuclear Regulatory Commission and Entergy point out that it would be very difficult for an airplane or attackers to cause a major release, and, in any case, security would be upgraded. Such assurances were not sufficient to allay public concern. In addition, concerns about accidents at or attacks on the spent fuel pools at Indian Point have been given new attention since 9/11 (Wald, 2005b). For instance, a National Research Council study (NRC, 2005) concluded that "successful terrorist attacks against spent fuel pools, although difficult, are possible"; the type of spent fuel pool at Indian Point, however, was not among those that the report considered most vulnerable. It should be noted that closing Indian Point would not by itself eliminate risk from the spent fuel, which may remain onsite for many years until a permanent storage disposal facility is ready.

In Westchester and surrounding counties, some 12 community groups (Hu, 2002) have called for the plant's closing (e.g., Riverkeeper, Public Citizen, and Indian Point Safe Energy Council).[11] Activities by these groups, including advertising and an HBO television special, have kept the issue of shutting down Indian Point on the political agenda. Riverkeeper claims that "a large radioactive release triggered by a terrorist attack on or accident at the facility could have devastating health and economic consequences. . . ." (Riverkeeper, 2006). Entergy, many safety analysts in the industry, and the Nuclear Regulatory Commission are convinced that a terrorist attack, even if it occurred, would be extremely unlikely to result in a large radioactive release. Riverkeeper also is concerned with environmental damage to the Hudson River, especially to fish, eggs, and larvae (van Suntum, 2005). Here, the policy issue, which is currently in the courts, is whether or not the river cooling system should be replaced by a more expensive system (Hu, 2003).

A key community concern has been the perceived inability of emergency plans to work in the aftermath of an accident or successful attack on the facility (Purdy, 2003; Lombardi, 2002). A state-sponsored study (Witt, 2003) found that "the plans do not consider the possible additional ramifications of a terrorist caused release." Early evacuation is not a requirement of Nuclear Regulatory Commission and state emergency planning because scenarios that would lead to early fatalities are not considered credible, even after 9/11. Yet the public appears to see early evacuation as crucial (Witt, 2003), which produces tension, because evacuation in the crowded New York metropolitan area is perceived by many to be impossible (Risinit, 2005). If many people attempted to evacuate or collect their families upon announcement of a potential release, the result could be gridlock (Witt, 2003; Westchester County, N.D.).

Local political leaders, such as Westchester County Executive Andrew Spano, call for an Indian Point shutdown, bringing the resources of the county to bear on the campaign. Rockland County Executive Scott Vanderhoef has also called for closure "before terror attacks" (Purdy, 2003). Congresswoman Nita Lowey, from New York's 18th District, has expressed concerns about the Indian Point facility and was responsible for commissioning this National Research Council study. She has also introduced a bill to require relicensed facilities to meet the same standards as those for new nuclear plants, which is currently not the requirement of the Nuclear Regulatory Commission.

As one indication of concern about reactor accidents, Westchester County, in cooperation with New York State, has developed a program to provide potassium iodide to residents who live, work, or travel within the 10-mile Emergency Planning Zone (Westchester County, N.D.). Such tablets, if taken early enough, significantly reduce radiation doses to the thyroid, the major risk after the Chernobyl accident.

In addition, Westchester County has commissioned expert studies on issues surrounding Indian Point (e.g., Levitan and Associates, 2005), as has Riverkeeper (Lyman, 2004; Komanoff, 2002; Schlissel and Biewald, 2002). The study for Westchester County highlighted the expense of an early shutdown of Indian Point, leading County Executive Spano to put his hopes on stopping Entergy in the relicensing process (Wald, 2005a).

Local opinion is by no means unanimous against Indian Point. Some political leaders are concerned that the plants have 1,200 employees and pay significant taxes to local schools and governments (Westchester County, 2003). Dan O'Neill, mayor of Buchanan, New York, home of the plant, is supportive of the facility (Purdy, 2003). Others are concerned over the reliability of the New York City power supply and potential increases in the costs of electricity.

CRITERIA FOR EVALUATING REPLACEMENT OPTIONS

The opportunities or options for replacing the Indian Point power plant are constrained by various technological, regulatory, and socioeconomic elements. These need to be taken into account in developing options for maintaining a reliable electric energy supply for southern New York State, while allowing for growth in the region.

Each of the constraints derives from somewhat different technological, regulatory, or cost considerations, many of which are unique to New York State. These constraints will affect both the choice and the timing of change in supply if Indian Point is considered for retirement.

For instance, the electricity supply available in New York currently relies heavily on Indian Point as a major baseload contributor to the power supply needed in the New York metropolitan area. Replacement of this capacity would require major efforts in new generation, transmission, and demand management.

Reliability of power supply depends on several factors,

---

[11] Information detailing these concerns can be found at the websites for the respective organizations, including www.riverkeeper.org, www.citizen.org, and www.ipsecinfo.org. Accessed March 2006.

including fuel availability, generation reserve, peaking load, and the growth rate of demand locally and in the region. Reliable electricity also hinges on the stability of the transmission-distribution system. In general, the NYCA system is carefully balanced to account for the location and operation of baseload plants, as well as intermediate and peaking units. Balancing is complicated by the nature of the generation, which includes not only conventional fossil and nuclear power sources but a variety of other technologies in the system, including hydroelectric units, wind power, and co-generated power at industrial facilities.

Safety has motivated this study to a great extent. Concern for public safety associated with a nuclear power plant close to the New York metropolitan area is substantial. However, there are additional considerations related to energy security and public safety. Security of the plant site must be maintained whether or not the plant is retired because it contains radioactive material, including stored spent fuel rods. Another energy security concern is fuel availability. In particular, most new generating units are fueled by natural gas, but gas supplies are limited and becoming increasingly expensive. Lengthy blackouts, whether caused by inadequate fuel supplies or transmission system instability, also threaten public health and safety. Imports of LNG may be required, but LNG also raises safety as well as energy security issues.

Adding to the complexity of decisions on closing Indian Point are issues of costs. Electricity costs are likely to rise if the area's low-cost power generator is retired. In addition, if the plant's lifetime is shortened, compensation to the owner may be required. Furthermore, the site will continue to require extensive security measures to protect the spent fuel until a more permanent storage facility is available. Costs are discussed in Chapters 4 and 5.

A complex web of environmental regulations must be considered with any alternative to the Indian Point plant. Regulations include national and local air and water quality and thermal discharge requirements as well as the possibility of constraints on greenhouse gas emissions associated with carbon fuel combustion. At the present time, air quality constraints are the most stringent for most alternative technologies. These are generally specified in terms of emissions of material regulated as criteria pollutants or hazardous air pollutants under the Clean Air Act (CAA) and its amendments and other requirements for airborne toxic chemical releases. New power plant sources are permitted only under very stringent constraints with regard to the CAA pollutants.

Finally, closing Indian Point and building new facilities, presumably at least partly elsewhere, would make significant differences in employment, tax base, and other community impacts. These changes might be positive or negative, but they must be included in the consideration of replacements for Indian Point.

Given the constraints corresponding to these criteria for the selection of options, the range of technologies available can be reduced substantially. It is unlikely that a 2,000-MW power plant would be built as an exact replacement for Indian Point, to be available just as Indian Point was closed. A package of demand and supply options, the latter possibly including new transmission lines as well as new generation, seems more plausible. The committee uses the following criteria to judge the proposed replacement packages for Indian Point:

1. Would the combination of demand and supply options provide adequate energy to replace that provided by Indian Point?
2. Would the generation and transmission system be adequate to deliver the energy reliably to end users?
3. How would the new combination of demand and supply options compare with Indian Point in terms of security of fuel supply for new generation?
4. How would economic costs, especially to the consumer, compare with continued operation of Indian Point?
5. How would environmental emissions and other impacts compare with continued operation of Indian Point?
6. What would be the impacts on local communities from closing Indian Point and replacing it with these options?

CONDUCT OF THE STUDY

This study was initiated by the U.S. Congress in the fiscal year 2004 Appropriations for the U.S. Department of Energy. The Committee on Alternatives to Indian Point for Meeting Energy Needs was formed in accordance with National Research Council procedures. The committee's statement of task is presented in the Preface. Biographical sketches of the committee members appear in Appendix A.

The committee held five full meetings over the course of the study. The first three meetings included open sessions at which many experts made presentations to the committee. The second meeting was held in White Plains, New York, to allow local residents interested in the issue to attend. Committee meetings and participants are listed in Appendix C. The project's website also invited viewers to submit comments.

In addition to the full committee meetings, several committee subgroups also conducted many conference calls and collectively prepared sections of this report.

The committee also contracted for two expert analyses. GE Energy built on its work with NYISO to analyze several scenarios for replacing the power from Indian Point. While NYISO generously allowed the committee to use its database, it should be noted that the scenarios were developed by the committee, not NYISO. Several members of the committee met in Schenectady, New York, to discuss scenarios and analytical methodology with NYISO and GE Energy, in preparation for the committee's analysis.

In addition, Optimal Energy of Bristol, Vermont, refined the 2003 analysis of energy efficiency potential that it had done for the New York State Energy Research and Development Authority to focus on the regions that would be impacted by the closure of Indian Point.

ORGANIZATION OF THE REPORT

There are two general options to consider in replacing Indian Point: reducing demand and increasing supply. As noted above, demand is increasing, but the growth rate can be controlled to some extent. Many efforts already are under way to increase the efficiency of use of electricity or to reduce demand during peaks when reliability concerns are highest. Chapter 2 discusses how those efforts could be expanded if it were necessary to compensate for the loss of Indian Point. It also discusses distributed generation and how that could affect load growth and electricity reliability.

Supply options, discussed in Chapter 3, include new generating units and transmission lines that can import power from underutilized generating plants in upstate New York and beyond. In recent years, almost all new generating plants have been fueled by natural gas, but those supplies are becoming strained. Modifying the bulk power system can be complicated, and many factors must be considered. In particular, reactive power has a large effect on transmission capability. The reactive power supplied by Indian Point would also have to be replaced if its units are closed.

Chapter 4 discusses institutional factors and various impacts that might result from the replacement of Indian Point with the options discussed in Chapters 2 and 3. Most new generating plants and transmission lines would be built by private companies, which could face daunting obstacles of regulation and financing. New facilities also would create a set of environmental impacts different from those created by Indian Point.

Chapter 5 analyzes several scenarios to evaluate the impact of closing Indian Point and replacing it with these other options. The scenarios with compensatory actions to replace Indian Point are to be viewed as representative of the actions that could be taken, not as a recommended path. Other combinations of options might prove less expensive or advantageous from other perspectives. Nor do these scenarios include all of the costs that could be involved, such as buying Indian Point in order to close it, or disposing of the spent fuel now being stored onsite.

A series of appendixes follow. Appendixes D through G, which give additional details on the options considered and the committee's analyses, are reproduced on the CD-ROM that contains the full report but are not included in the printed report owing to space limitations.

The committee's findings and conclusions are discussed in the Summary and Findings that precedes this chapter. This report does not include recommendations as to whether Indian Point should be closed.

REFERENCES

Hu, W. 2002. "Post-9/11, opposition to Indian Point plant grows." *New York Times*, April 24.

——. 2003. "Judge orders faster review of cooling unit at Indian Point." *New York Times*, April 10.

Komanoff, C. 2002. *Securing Power Through Energy Conservation and Efficiency in New York: Profiting from California's Experience.* Available at http://www.riverkeeper.org. Accessed April 2006.

Levitan and Associates. 2005. *Indian Point Retirement Options, Replacement Generation, Decommissioning/Spent Fuel Issues, and Local Economic/Rate Impacts.* Boston: Levitan.

Lombardi, K.S. 2002. "Indian Point and the Big If." *New York Times*, March 31.

Lyman, E. 2004. *Chernobyl on the Hudson? The Health and Economic Impacts of a Terrorist Attack at the Indian Point Nuclear Plant.* Washington, D.C.: Union of Concerned Scientists.

NEI (Nuclear Energy Institute). 2006. *U.S. Nuclear Power Power Plant Capacity, Capacity Factor and Generation.* Available at http://www.nei.org/documents/U.S.%20Nuclear%20Power%20Plant%20Capacity%20Capacity%20Factor%20and%20Generation.pdf. Accessed April 2006.

NRC (National Research Council). 2005. *Safety and Security of Commercial Spent Nuclear Fuel Storage: Public Report.* Washington, D.C.: The National Academies Press.

NYISO (New York Independent System Operator). 2005. *Comprehensive Reliability Planning Process (CRPP) and Reliability Needs Assessment.* Albany, N.Y. December 2005.

Purdy, M. 2003. "Our towns: Gospel of Armageddon finds fertile ground near Indian Point." *New York Times*, January 26.

Risinit, M. 2005. "Unlike Westchester, upstate Oswego welcomes nuclear power." *The Journal News.* Available at http://www.thejournalnews.com/apps/pbcs.dll/article?AID=/20050626/NEWS02/506260355/-1/spider/. Accessed June 26.

Riverkeeper. 2006. Available at http://www.riverkeeper.org. Accessed March 2006.

Schlissel, D., and B. Biewald. 2002. *The Impact of Retiring Indian Point on Electric System Reliability.* Cambridge, Mass.: Synapse Energy Economics.

van Suntum, L.R. 2005. "The cost of nuclear power" (Letter to the Editor). *New York Times*, May 23.

Wald, M. 1982. "Protests grow on Indian Point." *New York Times*, August 15.

——. 2005a. "County seeks deal on Indian Point, perhaps in vain." *New York Times*, June 19.

——. 2005b. "Study finds vulnerabilities in pools of spent nuclear fuel." *New York Times*, April 7.

Westchester County. N.D. "Emergency Planning for Indian Point." Available at http://www.westchestergov.com/indianpoint/. Accessed March 2006.

——. 2003. "Spano and Kaplowitz Announce Next Step in Effort to Replace Nuclear Energy at Indian Point." Available at http://www.westchestergov.com/currentnews/2003pr/IndianPointRFP.htm. Accessed April 2006.

Witt, J.L. 2003. "Review of Emergency Preparedness at Indian Point and Millstone" (Draft). Washington, D.C.: James Lee Witt Associates.

2

Demand-Side Options

DEMAND GROWTH IN THE INDIAN POINT SERVICE AREA

The New York Independent System Operator (NYISO) prepares compilations of historic electricity usage patterns and forecasts future electricity demand in New York State. Table 2-1 shows annual power consumption for selected years between 1993 and 2015 by region, in and around New York City and in the state, and Table 2-2 shows peak power requirements for the same years and areas. These consumption estimates are "weather-normalized" to enable comparisons across a typical year of weather (e.g., electricity use during years with particularly cold winters or hot summers was reduced to reflect what would have occurred during years with more typical numbers of heating and cooling degree-days).

Electricity use in the New York Control Area (NYCA) as a whole grew at about 1 percent annually between 1993 and 2004 as shown in Table 2-1. Demand in western New York and the Upper Hudson Valley actually dropped during that period. All of New York's demand growth has been downstate, with Long Island growing at 2.2 percent annually, New York City—even with the events of September 11, 2001—at 2.1 percent, and Zones H and I (most of the Lower Hudson Valley) at a rate of 1.4 percent.[1] This growth seems to be driven in part by a continuing expansion of the strong service sector (including government, education, and health care) that characterizes much of the downstate region. The manufacturing that once anchored the upstate economy has been in decline since the 1970s.

Summer peaks (Table 2-2), due largely to air conditioning, have grown more rapidly than has annual electricity use (Table 2-1), with Long Island seeing the highest growth in the state, followed by New York City and then the Lower Hudson Valley.

NYISO forecasts that the current growth rate in annual electricity use (though not that of peak-load growth) will continue out to 2015 in the Lower Hudson Valley, but with some slowing in New York City and Long Island (due to more limited opportunities for commercial and industrial expansion and greater investment in demand-management programs by Consolidated Edison). Consumption and peak load are forecast to grow at an approximately equal pace on Long Island and in New York City. Peak load is expected to grow slightly faster than consumption in the Lower Hudson Valley.

The projections of electricity demand in Tables 2-1 and 2-2 are predicated on the assumption that electricity prices will continue their historical decline, as shown in Figure 2-1. This assumption in turn depends on assumptions of fuel prices, generating mix, capital costs, and other factors. NYISO's demand forecasts are based on the relative trend in Figure 2-1, which was derived from analyses by the Energy Information Administration (EIA) for the Mid-Atlantic region (Energy Information Administration, 2006).

Such projections are highly uncertain for several reasons, most prominently:

1. Natural gas, which is the source of a large and increasing share of New York's electric generation, has shown large swings in price in recent years. Some of this has been temporary, for example owing to shortages in supply because of damage to equipment in the Gulf of Mexico region during the hurricanes of 2004 and 2005. More worrisome, however, has been the declining productivity of U.S. gas fields. The EIA expects gas prices to remain relatively stable over the next 10 years (Energy Information Administration, 2006). That may be the case, but probably only if imports of liquefied natural gas (LNG) are significantly increased. The only proposed LNG terminal in the state of New York, in Long Island Sound, faces vigorous opposition, as do other pro-

---

[1] The growth rates for Zones H and I alone appear to be higher than the overall rate for the Lower Hudson Valley, since a different NYISO report (*2004 Load and Capacity Data*, p. 7, Table I-4) shows no growth in Zone G.

TABLE 2-1 Weather-Normalized Annual Electricity Use, Past and Forecast, in Gigawatt-Hours per Year, for Three New York Regions and Statewide, Selected Years from 1993 Through 2015

Year	Lower Hudson Valley: NYCA Zones G, H, I[a]	New York City: NYCA Zone J	Long Island: NYCA Zone K	New York State: NYCA
1993	16,411	41,828	17,667	144,471
1997	16,206	44,676	18,185	148,008
2001	17,207	49,912	20,728	155,523
2005	19,625	52,836	23,178	164,050
2009	20,775	56,345	25,258	174,290
2013	22,610	58,949	26,598	180,710
2015	23,608	59,717	26,961	182,880
Growth per year:				
1993-2004	1.421%	2.071%	2.222%	1.004%
2004-2015	1.913%	1.194%	1.659%	1.151%

[a]NYCA, New York Control Area; Zone G, Hudson Valley; Zone H, Northern Westchester County; Zone I, rest of Westchester County.

SOURCE: Adapted from NYISO (2005a), p. 25.

TABLE 2-2 Weather-Normalized Summer Peak Power, Past and Forecast, in Megawatts, for Three New York Regions and Statewide, Selected Years from 1993 Through 2015

Year	Lower Hudson Valley: NYCA Zones G, H, I[a]	New York City: NYCA Zone J	Long Island: NYCA Zone K	New York State: NYCA
1993	3,337	8,365	3,595	27,000
1997	3,650	9,609	4,273	28,400
2001	4,421	10,424	4,901	30,780
2005	4,410	11,315	5,230	31,960
2009	4,849	11,965	5,580	33,770
2013	5,331	12,426	5,981	35,180
2015	5,590	12,648	6,112	35,670
Growth per year:				
1993-2004	2.365%	2.610%	3.270%	1.382%
2004-2015	2.380%	1.190%	1.618%	1.166%

[a]NYCA, New York Control Area; Zone G, Hudson Valley; Zone H, Northern Westchester County; Zone I, rest of Westchester County.

SOURCE: Adapted from NYISO (2005a), p. 26.

posed projects. Natural gas is discussed further in Chapter 3. If these supplies do not materialize, prices will rise and electricity costs will follow.

2. Even if the costs of production can be defined well, the wholesale price is a function of the auctions that NYISO conducts to procure supplies, as discussed in Chapters 1, 4, and 5. Price can be either above or below historic levels, depending on how many bidders are participating. The long-term impact of the New York process on prices to consumers is still uncertain.

Overall, if the price decline projected to start in 2006 does not occur, demand will be lower.

NYISO's new capacity-forecasting program is more rigorous than in the past, but even the best demand forecasts are not destiny. They are simply estimates, based on guesses about a host of parameters, which may prove to be too high or too low. Price increases, economic downturns, changes in fuel prices and availability, policy changes, and technological advance have all contributed to surprises in years past. Both in the 1970s and in late 1980s, serious power shortages were forecast for New York unless particular power plants

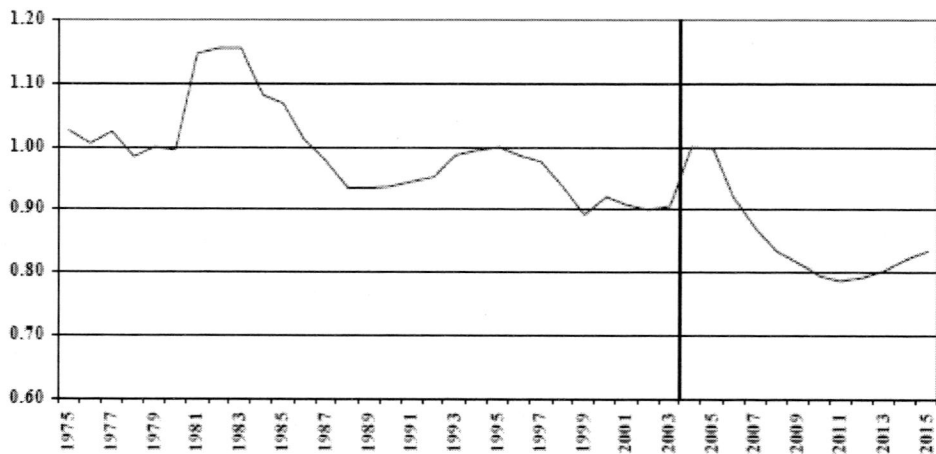

FIGURE 2-1 Past and projected trends in real residential electricity price in New York State relative to 1980. SOURCE: NYISO (2005a).

were built. Not all were, but no shortages occurred, and the demand for energy services was unfailingly met. The 1980s saga of Long Island's Shoreham nuclear plant, which was eventually closed before it produced any electricity, is one example. It is no criticism of the NYISO forecasts to observe that they do not reflect the full range of possibilities that could come into being if circumstances so required (such as an emergency shutdown of the Indian Point Energy Center or of another large generating source) or if state policies emphasized energy efficiency on the same scale as in California, as discussed later in this chapter.

The range of policy options available to power system operators and regulators has grown wider in recent years. It now includes energy efficiency, load management, integrated resource planning, and performance-based rate making with incentives for cost-effective energy efficiency.

New York State's spending on efficiency in the electric sector declined significantly in the mid-1990s, falling from a peak of some $300 million per year in the early 1990s to a low of some $50 million per year in 1996. The state's only performance-based rate-making plan based on capping revenues[2] lapsed in 1997. The New York State Energy Research and Development Authority (NYSERDA) now spends about $150 million annually on energy efficiency programs, discussed below (NYSERDA, 2005b). Comparing trends in consumption and peak load between 1993 and 1997 with those between 1997 and 2001 (Tables 2-1 and 2-2) suggests that the demand-side management (DSM) program cutbacks may have allowed demand to grow faster than it would have with stronger programs.

POTENTIAL OF DEMAND-SIDE OPTIONS

The impacts of current and planned programs for reducing electricity consumption and peak electrical loads could be among the most cost-effective replacements for the energy provided by the Indian Point Energy Center. This section describes promising demand-side control options, including estimates of their achievable potential and barriers to their implementation. The focus is on the ability of demand-side options to reduce on-peak requirements of consumers for electricity. While Indian Point is a baseload plant, the biggest challenge to replacing its capacity occurs during summer and winter peaks when regional generating resources and transmission capacity are most constrained—hence the focus on demand-side options that could displace peak loads. The ability of energy efficiency to reduce megawatt-hours of electricity consumption and levels of consumer bills in the residential and commercial sectors is highlighted in Appendix G-1 ("Demand Reduction").

---

[2]Revenue-cap plans are more compatible with energy efficiency than are the more common price-cap plans because they adjust revenues to avoid any loss in profitability arising from declining sales. Cost-effective energy efficiency can lower bills while raising prices (because the decline in consumption more than offsets the increase in prices).

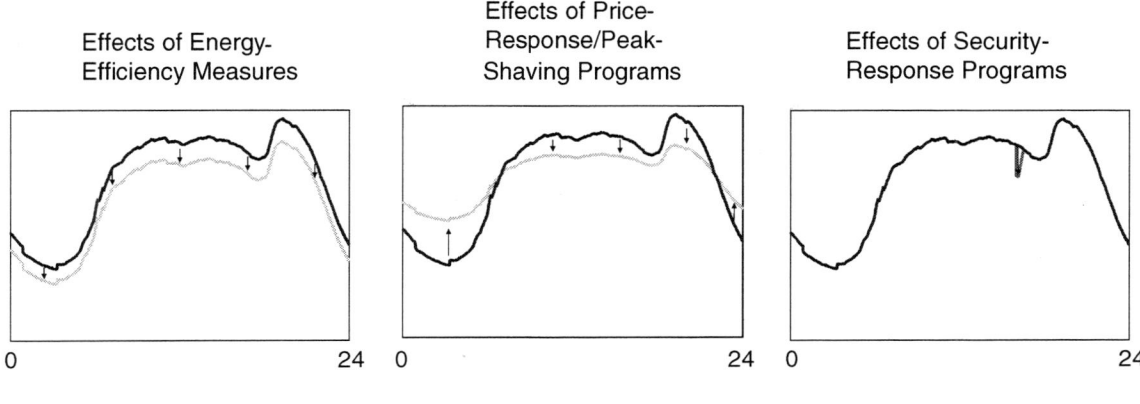

FIGURE 2-2 Effects of demand-reduction programs on daily power demand. SOURCE: Adapted from Kirby et al. (2005); Gillingham et al. (2004).

Definition of Demand-Side Options and Measures of Potential

Demand-Side Options

This chapter considers two types of demand-side options:

- *Energy efficiency programs* (principally in the commercial and residential sectors) and *demand-response (DR) programs* (including permanent and "callable" resources), and
- *Distributed generation (DG)*, which is generally not dispatchable and thus not included in most electrical system reliability analyses. DG includes combined heat and power (CHP) systems and distributed photovoltaics (PV).

Energy Efficiency and Demand-Response Programs. Energy efficiency programs allow users to perform the same functions that they normally would, but with less energy consumption. When applied to electricity uses, improved efficiency reduces demand throughout the day, often with the greatest effect during peak demand. The left panel of Figure 2-2 shows a typical daily cycle of demand, low at night, rising during the day, and peaking during the late afternoon. The lower curve shows demand with improved efficiency of use. Energy-efficiency improvements can be expensive, but once implemented they can save energy for many years. Reductions in peak-power requirements can also contribute to system stability in the event of sudden disturbances such as a loss of system components or short circuits.[3] Furthermore, reducing peak demand means that generating capacity and reserve margins can both be reduced. Thus, investments in reducing peak demand through energy efficiency measures can have a value of 118 percent of the actual reduction in avoiding the addition of new capacity.[4]

Energy-efficiency mechanisms can include mandatory efficiency standards for buildings and appliances; targeted financial incentives and assistance; codes; information and education programs; and research on energy-efficient technologies (Silva, 2001, pp. 96-104; Brown et al., 2005, pp. 45-60). They can take place in a variety of program areas, including residential lighting; single-family weatherization; nonresidential heating, ventilating, and air conditioning (HVAC); and new construction (National Energy Efficiency Best Practices Study, 2004). Stimulating greater investments in energy-efficiency measures is complex, however, since it involves multiple actors and agents, including varied consumers, vendors, independently owned utilities, unaffiliated distribution companies, and federal, state, and local agencies (Harrington and Murray, 2003).

One well-documented stimulant for energy efficiency is that of increased electricity prices. Most models of electricity markets incorporate an estimate of the price elasticity of demand for electricity. Consistent with past research, one recent study of price response based on 119 customers from New York State (Goldman et al., 2005) confirms that customers' price response is generally modest. In particular, the surveyed customers had an average price elasticity of 0.11, which means that their combined ratio of peak to off-peak electricity usage declines by 11 percent in response to a doubling of peak prices (relative to off-peak prices). Thus, price

---

[3]The adequacy and security aspects of electrical system reliability are briefly discussed in NYISO's report *Reliability Needs Assessment* (NYISO, 2005a).

[4]The North American Electric Reliability Council has set a standard of 18 percent for reserve generation. This criterion has been adopted by the New York State Reliability Council.

increases in the event of more-constrained supplies could produce a measurable reduction in demand, but the overall effect would be modest in magnitude. While long-term price elasticities of demand are likely to be larger, their impact would occur outside the time frame of interest for this report.

Demand-response programs focus on consumers' actions to change the utility's load profile. These programs are not aimed at saving energy so much as at shifting the time at which it is demanded, as shown in the middle set of curves in Figure 2-2 (Gillingham et al., 2004). Price-response programs move consumption from day to night or curtail discretionary usage. Peak-shaving programs focus on reducing peaks on high-load days by requiring greater response during peak hours. These programs allow utilities to better match electrical demand with their generating and transmission capacity. By changing the load curve for utilities, system reliability can be enhanced and new power plant construction can be avoided or delayed. Overall costs are reduced because peak power is more expensive than average costs.

Demand-response programs allow consumers to respond to electricity prices directly, offering mechanisms to help manage the electricity load in times of peak electricity demand in order to improve market efficiency, increase reliability, and relieve grid congestion. Significant consumer benefits can also accrue from real-time demand-response programs, chiefly in the form of cost savings due to lower peak electricity prices, less opportunity for market manipulation by electricity providers, and additional financial incentives to induce consumer participation in these programs.

Security-response programs enable utilities to drop loads in response to electric system contingencies. These programs can be implemented quickly and inexpensively, usually with the agreement of large users of electricity, who receive lower rates in return for relying on interruptible power. These programs have no impact on the load except during peak periods, as shown in the right-hand panel of Figure 2-2.

Distributed Generation. Distributed generation is the production of electricity at or close to its point of use. DG technologies include internal combustion engines, fuel cells, gas turbines and microturbines, Stirling engines, hydro, and microhydro applications, photovoltaics, wind energy, solar energy, and waste and biomass fuel sources. DG is usually installed on the customer side of the meter and is not dispatchable by the utility. DG ranges in size from a few kilowatts (kW) to 20 or even 50 megawatts (MW). Recent manufacturer interest and sales growth have been particularly strong in the 50 kW to 5 MW range. An objective has also been to move away from traditional diesel generators, up to now a common but relatively "dirty" source of distributed generation.

Combined heat and power, a subset of DG, generally involves reciprocating engines or turbines to drive electric generators, with the waste heat captured and used for other purposes. Typically, CHP systems generate hot water or steam from the recovered waste heat and use it for process or space heating. The heat can also be directed to an absorption chiller where it can provide process or space cooling. CHP systems may offer economic benefits, security, and reliability.

Siting generation close to its point of use, as with CHP systems, enables greater use of a device's overall energy output. Historically the average efficiency of central-station power plant systems in the United States has been approximately 33 percent, and until quite recently had remained virtually unchanged for 40 years. This means that about two-thirds of the energy in the fuel cannot be converted to electricity at most power plants in the United States and is released to the environment as low-temperature heat. CHP systems, by capturing and converting waste heat, achieve effective electrical efficiencies of 50 to 80 percent. Furthermore, centrally located facilities typically lose 5 to 8 percent of their rated output through transmission and distribution losses.[5] CHP systems, by being at or near the point of use, avoid most of these losses.

The improvement in efficiency provided by combined heat and power reduces emissions of carbon dioxide and usually other air pollutants. Since CHP requires less fuel for a given energy output, it reduces the demand for key fuels such as natural gas, coal, and uranium.[6] CHP can help reduce congestion on the electric grid by removing or reducing load in areas of high demand and can also help decrease the impact of grid power outages. NYSERDA comments that "energy savings [from CHP systems] represent a social benefit in lowering the pressure on fuel and electricity supply and infrastructure, thereby providing lower prices for all consumers."[7] Mayor Michael Bloomberg's New York City Energy Task Force, in considering options to reduce electrical capacity problems in the city, concluded that "distributed resources can reduce or reshape electric system load and thereby mitigate the need for increased generation and/or transmission resources. . . . With appropriate policies and incentives, distributed resources are often the most readily available, cost-effective, and underutilized clean energy resources that can potentially reduce or defer the amount of required new electric supply from generation and transmission systems. While it can take many years to plan, design and build electric generation plants, most distributed resources can be deployed within a year."[8] A dispersed network of DG units is also less vulnerable to terrorism, whether from direct attacks or computer hacking, than a single large power station.

---

[5] Available at http://www.epa.gov/chp/what_is_chp/why_epa_supports_chp.htm. Accessed October 3, 2005.

[6] Available at http://www.epa.gov/chp/what_is_chp/benefits.htm. Accessed October 3, 2005.

[7] Available at http://www.nyserda.org/programs/pdfs/CHPFinalReport2002WEB.pdf. Accessed October 3, 2005.

[8] Available at http://www.nyc.gov/html/om/pdf/energy_task_force.pdf. Accessed October 3, 2005.

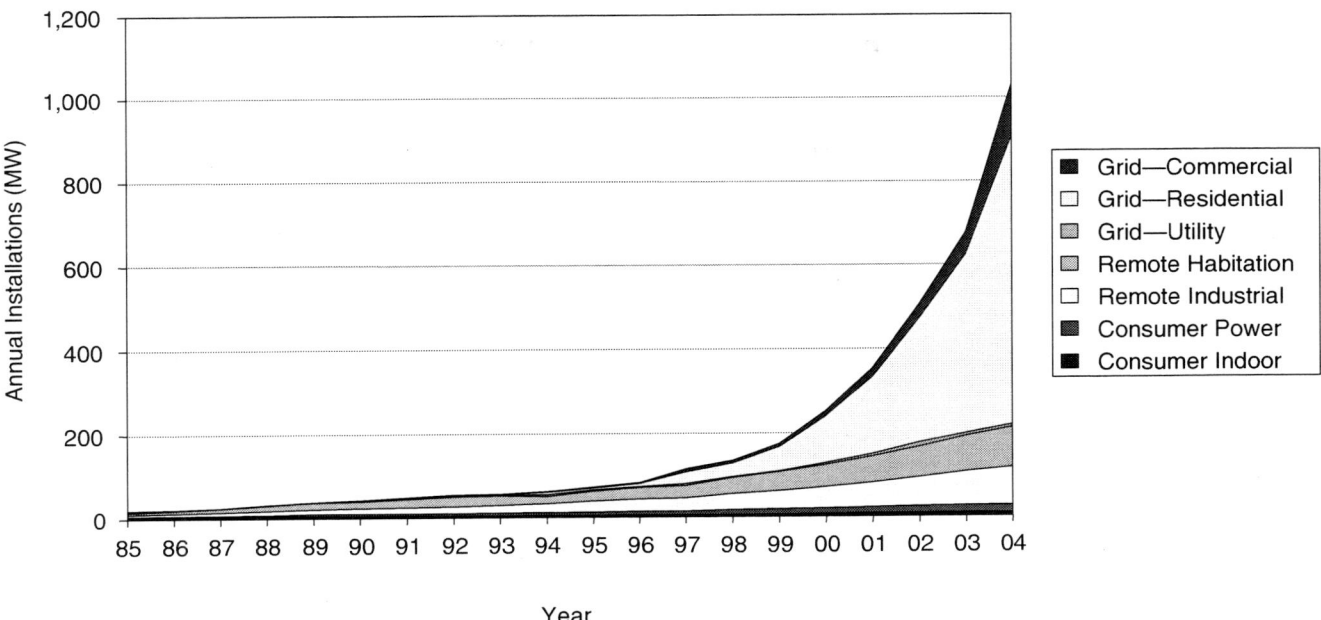

FIGURE 2-3 Global photovoltaic market evolution, by market segment, 1985 to 2004 (42 percent average annual growth). SOURCE: Personal communication from Paula Mints, Senior Photovoltaic Analyst, Strategies Unlimited, Mountain View, Calif., February 11, 2005.

Photovoltaic (PV) technology generates electricity from sunlight in a system with no moving parts. PV units can be mounted on rooftops and left largely untended. This DG option, when installed for the end user, competes against retail, not wholesale, electricity rates. Since its production profile is nearly coincident with the summer peak demand, it can contribute significantly to grid stability, reliability, and security. Thus, from a planning perspective PV should be valued at a rate closer to the peak power rate than the average retail rate.[9] The cost of PV-generated electricity is expected to decline considerably over the next decade, falling from a current cost of 20 to 40 cents per kilowatt-hour (¢/kWh) to a projected cost of 10 to 20 ¢/kWh by 2016, less than the retail price of electricity in New York City (USDOE, 2004; Margolis and Wood, 2004; SEIA, 2004).[10] Thus, PV may be in the economic interests of New York customers sooner than others in sunnier parts of the country.

Growth of the global PV market from 1999 to 2004 has averaged 42 percent annually (see Figure 2-3). Large-scale production will contribute greatly to continuing cost declines. As shown in Figure 2-3, the fastest growth was in the grid-connected residential and commercial segments.

Measures of Potential

When evaluating the potential for additional demand-side options to be deployed in future years, four types of estimates are generally used.

- *Technical potential* refers to the complete penetration of all applications that are technically feasible.
- *Economic potential* is defined as that portion of the technical potential that is judged cost-effective.
- *Maximum achievable potential* is defined as the amount of economic potential achievable over time under the most aggressive program scenario possible. It takes into account administrative and program costs as well as market barriers that prevent 100 percent market penetration.
- *Program potential* is the amount of penetration that would occur in response to specific program funding measures (Rufo and Coito, 2002; NYSERDA, 2003).

Current Programs Operating in the Indian Point Territory

When assessing the additional potential for demand-side options in the Indian Point service territory, it is necessary to

---

[9]PV power replaces power that the home owner or business owner would have had to buy from the grid. Therefore, its value is at the retail level. PV power usually peaks around midday, when sunlight is strongest. Air conditioning loads peak several hours later as buildings heat up, but a PV system would still be putting out a high fraction of its peak output at that time of day.

[10]There is wide variation in retail rates across New York State, but a New York City resident may pay over 20 ¢/kWh. See http://www.dps.state.ny.us/bills.htm. Accessed March 2006. Commercial and industrial customers would pay less for larger quantities.

TABLE 2-3 Current Photovoltaic (PV)-Related Policies in New York State

Incentive	Description
Sales tax exemption (R)	100% sales tax exemption.
Property tax exemption (C, I, R, A)	15-year tax exemption for all solar improvements.
Personal tax credit (R)	25% tax credit for PV (<10 kW) and solar hot water (SHW), capped at $5,000.
State loan program (C, I, R, A, G)	$20,000 to $1 million loan for 10 years at 4 to 6.5% below the lender rate for PV and SHW.
State rebate program (C, I, R, A, G)	$4 to $4.50/W (<50 kW) up to 60% of total installed costs. Investor owned utilities' customers only.
Municipal utility rebate program (C, R, G)	$4 to $5/W (<10 kW). Long Island Power Authority customers only.
Interconnection standards (C, I, R, A)	Standard agreement for PV requires additional insurance and an external disconnect. Up to 2 MW maximum.
Net metering standards (R, A)	All utilities must credit customer monthly at the retail rate for PV systems under 10 kW.

NOTE: C = commercial, R = residential, I = industrial, A = agricultural, G = government.

SOURCE: Incentive data available at www.DSIRE.org. Accessed April 21, 2006.

characterize the programs that are currently in place and the results achieved to date. NYSERDA is spending a total of $1.2 billion (or $175 million annually over a 7-year period) in public and private funds in the state of New York (NYSERDA, 2005a, p. ES-7). NYSERDA estimates that its programs have reduced peak demand by 860 MW and reduced electricity consumption by 1,400 gigawatt-hours (GWh) annually. At a delivered price of about $0.03/kWh, NYSERDA estimates that the technical potential for its efficiency programs in New York State is 20,000 GWh and a cumulative 3,800 MW reduction of peak load by 2012, with corresponding forecasts for 2022 of 41,000 GWh and 7,400 MW.[11]

New York State's 2002 State Energy Plan sets forth "the goal of becoming a national leader in the deployment of distributed generation technology" and recommends that the state "should take all reasonable steps necessary to facilitate the interconnection of DG and CHP resources into the electricity system and increase the use of DG and CHP resources in the State."[12]

Progress has been made on several fronts over the past several years in advancing combined heat and power systems in the United States. The Bush administration promoted CHP in its National Energy Plan, and the Energy Policy Act of 2005 directs states to consider adopting interconnection standards for CHP and to promote the development of CHP technologies. National model emissions regulations are under development by several organizations, and the Federal Energy Regulatory Commission (FERC) has issued small-generator interconnection standards as well as a model state rule.

Many states and regions are conducting their own rulemaking processes on interconnection policies, emissions barriers, and tax issues for CHP. Most relevantly, the New York Public Service Commission has both reduced the standby electricity rate charges for CHP and set up an attractive natural gas rate structure for CHP. Both of these actions apply in the Consolidated Edison service territory. New York State, through NYSERDA, also has the largest incentive program for CHP in the nation.

New York also has enacted policies aimed at encouraging the adoption of photovoltaic technology, as shown in Table 2-3. The result is a comprehensive set of incentives for residents and businesses to install PV. The incentives take the form of tax exemptions and credits, loan subsidies, rebates (administered by the Long Island Power Authority and NYSERDA), and standard interconnection and metering rules that are exceeded in the Northeast only by New Jersey.

New York's existing rebate or "buy-down" program is administered by NYSERDA. It is called New York Energy $mart and includes customers of all major investor-owned utilities. New York Energy $mart provides customers who purchase and install PV systems with a $4 per watt rebate. This incentive, in combination with state tax credits and exemptions, has resulted in the installation of more than 1.5 MW by the summer of 2005. The program currently has $12 million allocated to its PV incentive program, of which about $6.5 million has been reserved as installer/customer incentives. The remaining funding should take the program through 2006.

The following subsections describe the energy-efficiency, demand-response, and distributed-generation programs that are in operation or planned for implementation in the near future by the three major power providers in downstate New York: Consolidated Edison (ConEd), the New York Power Authority (NYPA), and the Long Island Power Authority (LIPA).

---

[11]Paul A. DeCotis, NYSERDA, 2005. "New York State's Public Benefits Energy Efficiency Programs," presentation to the National Research Council Committee on Alternatives to Indian Point for Meeting Energy Needs, Washington, D.C., June 1, p. 5.

[12]Available at http://www.nyserda.org/sep/sepsection1-3.pdf. Accessed October 3, 2005.

Consolidated Edison

Consolidated Edison has established demand-management subsidy programs as follows (Plunkett and Gupta, 2004):

- Overarching goal: Reduce projected peak-load growth by 535 MW through demand management.
- NYSERDA Systems Benefit Charge (SBC) II programs: 250 MW (80 MW permanent) in ConEd service territory (already accomplished).
- NYSERDA SBC III programs: 300 MW (120 MW permanent) in ConEd service territory.
- "Incremental" programs to provide 300 MW of peak-load reduction, including the following:
  —ConEd: up to 150 MW in constrained networks.
  —NYSERDA: up to 150 MW throughout ConEd's service territory (after accomplishing the 550 MW in SBC II and III). Budget is $112 million.

The following measures are being emphasized in NYSERDA's incremental programs:

- Energy efficiency (goal of 68 MW)—Commercial and Industrial Performance Program (CIPP), New Construction, Smart Equipment Choices, Energy $mart Loan Fund, Building Performance Program, Flexible Technical Assistance.
- Load management (goal of 55 MW)—Peak Load Reduction and Aggregated Load Reduction programs.
- Distributed generation (goal of 27 MW)—Clean DG Incentives Program for engines and microturbines.

New York Power Authority

The following energy services programs are operated or planned by the New York Power Authority:

- NYPA has committed $100 million a year for energy-efficiency projects through performance contracting with its private- and public-sector customers.
  —Cumulative reductions for 1987 through 2004 were 900 GWh and 194 MW.
  —Cumulative estimated emissions reductions were approximately 491,000 tons of CO_2; 1,350 tons of SO_2; and 675 tons of NO_x.
- NYPA materials state that 1,200 energy-efficiency projects have taken place at approximately 2,200 public buildings across New York State.
- Measures through NYPA's energy services programs are primarily lighting, motors, and HVAC and limited to a maximum payback period of 10 years.

NYPA also has established three renewable resources projects, including the following:

- Nine fuel cell installations totaling 2.4 MW using waste gas produced from sewage plants.
- 18 rooftop photovoltaic systems with a combined capacity of 570 kW.
- As of December 31, 2004, 4 million electric-drive vehicle miles for hybrid-electric transit buses, all-electric school buses, station commuter cars, electric delivery trucks, electric low-speed vehicles, and other technologies.

Long Island Power Authority

Beginning in May 1999, LIPA committed $355 million over 10 years for energy-efficiency projects, clean distributed generation, and renewable technologies. Through the end of 2004, LIPA had spent approximately $170 million, or approximately $34 million a year. This Clean Energy Initiative is estimated by LIPA to have had the following impacts:

- Annual savings are estimated at 330 GWh, with 326 MW of permanent demand reductions and 145 MW of curtailable demand reduction.
- Annual emissions reductions are approximately 1,400 tons of SO_2; 500 tons of NO_x; and 355,000 tons of CO_2.
- Through the first 5 years of deployment, cumulative emissions reductions are estimated at 1.3 million tons of CO_2; 1,900 tons of NO_x; and 5,000 tons of SO_2.
- LIPA estimates that approximately 3,500 "secondary" jobs have been created as a result of the program.

The Clean Energy Initiative includes the following kinds of programs:

- *Residential*—lighting and appliances, HVAC, and the Residential Energy Affordability Program (REAP), which provides free installation of efficiency measures and education for low-income households. In addition, LIPA launched the Solar Pioneer Program for photovoltaics in 1999, offering customers a substantial rebate. The rebate's budget is tied to LIPA's 5-year Clean Energy Initiative, with funding totaling $37 million annually (covering multiple technologies). The Clean Energy Initiative is expected to receive funding through 2008. To date, 511 rebates have been disbursed for PV systems totaling more than 2.63 MW installed on Long Island. LIPA's rebate is currently set at $4/W.
- *Commercial and industrial*—commercial construction and peak reduction programs.
- *General*—the Customer-Driven Efficiency Program, providing custom assistance for residential and commercial customers; LIPA*edge*, a direct load-control program.
- *Research and development*—wind power, fuel cells, electric vehicles, hybrid-electric buses, tidal power, wave power, geothermal, and various electrotechnologies.
- *New York ENERGY STAR Labeled Homes Program*—introduced by LIPA with NYSERDA in July 2004.

The Potential for Additional Energy-Efficiency Improvements

The preceding review shows that New York State is reaping substantial gains from its programs for reducing electricity consumption. In fact, NYISO projects that the growth rate of consumption for the New York City area will be lower than in the recent past, in part because of these activities by NYSERDA, ConEd, NYPA, and LIPA. This subsection estimates the potential for further gains if these programs are expanded.

Targets for Additional Energy-Efficiency Improvements

One study (NYSERDA, 2003) estimates the potential for energy-efficiency improvements in New York State and provides details for Zones J (New York City) and K (Long Island outside of New York City). The study focuses on 3 years—2007, 2012, and 2022—and analyzes residential, commercial, and industrial sectors separately. The study is based on detailed information about technologies (e.g., 87 technologies or technology bundles for commercial buildings). It concludes that most of the economic potential for energy-efficiency improvements is concentrated in the commercial and residential sectors and not in the industrial sector.

For instance, NYSERDA (2003) forecasts that 3,726 GWh of economic potential would exist by 2007 in the residential sector of New York City, and that this would grow to 4,461 GWh by 2012. The residential efficiency measures that hold the most promise include the following:

- *Lighting*—compact fluorescent lightbulbs, fluorescent light fixtures, outdoor light controls, light-emitting diode (LED) nightlights, ceiling fans with fluorescent lights, multifamily common areas with specular reflectors, motion sensors, and LED exit signs;
- *Cooling*—efficient central air conditioners, air source heat pumps, ground source heat pumps, duct sealing, duct insulation, room air conditioners, humidifiers, new-construction HVAC systems;
- *Refrigerators*—upgrades to more efficient refrigerators, removal of second refrigerators or freezers;
- *Electronics*—computer monitors and central processing units (CPUs), laser printers, fax machines, exhaust fans, power supply, waterbed mattress pads, and waterbed replacement; efficient clothes washers; efficient televisions, VCRs, and DVD players;
- *Space heating*—efficient furnace fans, programmable thermostats, ENERGY STAR windows, blower door guided air-sealing, attic insulation, wall insulation, foundation insulation, heating controls, heat-recovery ventilators, and improved baseboard systems; and
- *Domestic hot water*—upgrade of heat-pump water heaters, upgrade of efficient well pumps, wastewater heat recovery, hot-water conservation measures, desuperheater off-ground source heat pumps.

In the commercial sector of New York City, NYSERDA (2003) forecast that 12,567 GWh of economic potential would exist by 2007 and that this would grow to 13,712 GWh by 2012. The commercial efficiency measures that hold the most promise include these:

- *Indoor lighting*—lamp ballasts, fixtures, specular reflectors, compact fluorescent lightbulbs, high-efficiency metal halides, occupancy sensors controls, daylight dimming, LED exit signs;
- *Refrigeration*—high-efficiency vending machines, vending misers, high-efficiency refrigerators, high-efficiency reach-in coolers, high-efficiency ice makers, walk-in refrigeration retrofit package, heat pump water heater;
- *Cooling*—high-efficiency air conditioning, high-efficiency heat pumps, high-efficiency chillers, optimized HVAC systems, optimized chiller distribution and control systems, water source heat pump, ground source heat pump, emergency control, dual enthalpy control, high-efficiency stove hoods, high-performance glazing;
- *Ventilation*—emergency management system control, premium efficiency motor, variable-frequency drive;
- *Office equipment*—high-efficiency CPUs, high-efficiency monitors, low-mass copiers, high-efficiency fax machines, high-efficiency printers, high-efficiency internal power supplies;
- *Whole-building controls*—retrocommissioing, commissioning, integrated building design, high-efficiency transformers;
- *Water heating*—high-efficiency tank-type water heater, point-of-use water heater, booster water heater, heat pump water heater;
- *Outdoor lighting*—LED traffic lights, LED pedestrian signs, pulse-start metal halides, compact fluorescent bulbs, improved exterior lighting design;
- *Space heating*—high-efficiency heat pumps, water source heat pumps, ground source heat pumps, optimized HVAC systems, optimized chiller control systems, emergency management control systems, high-efficiency stove hood, high-performance glazing; and
- *Miscellaneous*—high-efficiency clothes washers, water and wastewater optimization.

A more detailed account of the potential for these measures appears in Appendix G-1.

NYSERDA's $175 million New York Energy $mart Program (funded by New York's Systems Benefit Charge program, through a surcharge to each consumer's bill) has shown that efficiency programs can be successful. A 2004 evaluation of New York Energy $mart concluded that five efficiency programs have saved around 1,000 GWh from 2003 through 2004. The same review concluded that full implementation of New York Energy $mart is expected to achieve 2,700 GWh in the next 2 years.

These programs already are accounted for in the NYISO demand projections. Expanding current programs and creating new ones could achieve further gains in efficiency. If Indian Point is to be closed, that is one of the replacement options that can be considered.

Potential for Peak-Demand Reduction

Energy-efficiency programs can save considerable electricity, and the NYSERDA (2003) study documented that a great many improvements are available at modest cost. However, not all improvements will save at the same moment. The key consideration in the possible replacement of Indian Point is that of maintaining reliability during periods of peak load. By lowering overall demand, energy-efficiency programs also reduce peak demand, but not by the total of all the improvements.

The committee estimated the peak-load reduction that might realistically be achieved as a result of efficiency programs in the Indian Point region, as shown in Table 2-4. Details of the estimation are provided in Appendix G-2, "Estimating the Potential for Energy-Efficiency Improvements."

It is unlikely that programs can be put in place with sufficient resources to deliver all of the maximum achievable potential. The program potential is estimated at half the achievable potential. This factor is intended to introduce additional conservatism into estimates of the potential for energy efficiency. It is consistent with the estimate of Rufo and Coito (2002, Table 3-3) of the lower bound for advanced efficiency in California at one-half the higher bound for maximum achievable efficiency. The application of this factor results in estimates for program potential that grow from a reduction of 420 MW in 2007 to a reduction of 550 MW in 2015.

Two final adjustments are shown in the bottom line of Table 2-4. First, some lead time is required to phase in and establish new programs and expand existing activities. Programs established or expanded in 2006 will have very limited effect in 2007. Therefore, the program potential of 420 MW in 2007 is reduced to a phased-in programmable potential of 100 MW. The phased-in programmable potential is assumed to grow rapidly to 450 MW in 2010 and to reach the level of the full program potential of 550 MW by 2015. In addition, the committee expects that high fuel prices will increase the incentive to improve efficiency. Therefore the estimated phased-in programmable potential in 2015 is increased to 575 MW.

The estimates in Table 2-4 are consistent with those of other studies. The New York Energy $mart review noted above expected a reduction of peak demand of 880 MW within 2 years (statewide) as a result of program activities. A study presented to the New York State Public Service Commission concluded that the achievable potential for efficiency measures in New York City was 283 MW for residential and 1,392 MW for commercial buildings over 10 years (Plunkett and Gupta, 2004).

Finally, a study of the energy-efficiency potential in the New York City area, sponsored by the Pace Law School Energy Project and the Natural Resources Defense Council, concluded that savings of 1,163 MW to 3,032 MW peak demand could be achieved by aggressive energy-efficiency programs within 2 years (Komanoff, 2002).[13] To accomplish such reductions, the study suggested applying the rapid "crash efficiency" techniques—targeting the deployment of more efficient lighting, air conditioners, and appliance standards—employed by the state of California after its energy crisis in 2001. The extreme conditions associated with California's 2001 programs are not the context within which options for Indian Point are being evaluated, but they do illustrate a higher bound of possibilities if energy efficiency were to become a political rallying cry in New York City.

The Potential for Future Demand Response

Several of NYSERDA's existing programs illustrate the ability of demand-response programs to reduce peak electri-

TABLE 2-4 Committee Estimation of the Potential of Energy-Efficiency Programs in New York Control Area Zones I, J, and K, Selected Years Between 2007 and 2015 (MW)

Maximum Achievable Potential	Reductions in Year				
	2007	2008	2010	2013	2015
Zone I (Westchester County)	113	119	127	140	148
Zone J (New York City)	502	529	563	624	658
Zone K (Long Island outside of New York City)	226	239	253	285	297
Total maximum achievable potential	842	887	943	1,046	1,103
Total program potential (50% of achievable)	420	440	470	520	550
Phased-in programmable potential	100	200	450	525[a]	575[a]

NOTE: Details of the estimation are provided in Appendix G-2, "Estimating the Potential for Energy-Efficiency Improvements."

[a]Note that the "phased-in programmable" estimates exceed the "total program potential" in these years. This reflects the fact that more efficiency investments are cost-effective with the increased price of fuels today, and this is likely to be the case well into the future. These figures are based on historic (and low, by today's standards) Energy Information Administration price forecasts to calculate cost-effective energy efficiency.

SOURCE: Derived from NYSERDA (2003).

---

[13]This "lowest" estimate included adjustments for climate, forecast uncertainties, and consumption patterns.

cal loads for costs per kilowatt that are far lower than the cost of installing new peak capacity. Three of these programs alone have already avoided the need for over 700 MW of peak capacity:

- *Peak Load Reduction Program:* avoids the need for between 355 and 375 MW,
- *Enabling Technology for Price Sensitive Load Management Program:* avoids the need for 308 MW, and
- *Keep Cool Program:* avoids the need for between 38 and 45 MW.

NYSERDA divides its efficiency programs into three types: business/institutional (which include the Commercial and Industrial Performance Program, New Construction Program, and Peak Load Reduction Program); residential (which includes the Keep Cool Program); and low-income (which includes the Low-Income Assisted Multi-Family Program).[14]

In the studies referred to here, the prices reflect capacity costs and expenses for the downstate and urban areas. The analyses use avoided costs based on wholesale-electricity bid prices (rather than production costs), and they use energy-efficiency load profiles to differentiate savings by time of day (NYSERDA, 2004b, p. 1).

The studies evaluating NYSERDA programs also distinguish between proposed megawatts (demand target), enabled megawatts (coincident demand reduction), pledged megawatts (based on self-reporting), and delivered megawatts (averaged hourly reduction). Most of the estimates below (unless otherwise noted) refer to pledged megawatts. When some of the evaluations listed the delivered megawatts, they were typically only half the pledged rate. On the other hand, the estimated cost per megawatt of demand reduction is generally much lower than that of new supply options.

Peak Load Reduction Program

The Peak Load Reduction Program (PLRP), created in 2000, uses four different program segments:

1. *Permanent demand-reduction efforts*, which result in reduced demand through the installation of peak-demand-reduction equipment;
2. *Load curtailment and shifting*, through enrollment in the NYISO demand-response program;
3. *Dispatchable emergency generator initiatives*, which allow owners of backup generators to remove their load from the grid in response to NYISO requests; and
4. *Interval meters*, which reduce peak demand at the site of consumption.

The program avoids between 355 and 375 MW of peak demand. However, 340 MW of this is "callable," and only about 15 to 20 MW are permanent. Participants that are callable receive annual capacity payments and are required to perform when called. The program costs around $42.7 million over 8 years, or approximately $120/kW of peak-load reduction.

Enabling Technologies Program

The Enabling Technologies Program (ETP), created in 2000, supports innovative technologies that enhance load serving entities (LSEs), curtailment service providers (CSPs), and NYISO. It directs customers to reduce load in response to emergency or market-based price signals. The technologies used include advanced meters, transaction-management software, and networking and communication solutions. As of 2003, the ETP has saved 308 enabled peak MW. The program costs around $34.4 million per 8 years, or approximately $110/kW of peak-load reduction.[15]

Together, the PLRP and ETP saved 174 MW in 2001, 311 MW in 2002, and 288 MW in 2003.[16]

Keep Cool Program

The Keep Cool Program was started in 2001 and ended in 2003. It encouraged the replacement of old, inefficient air conditioners with new ENERGY STAR-rated room air conditioners and through-the-wall units. The program has two main components: it includes rebates and incentives for customers, and it uses a significant marketing campaign that encourages customers to shift appliance use to nonpeak periods. As a result of the wide scope of its multimedia marketing program, the Keep Cool Program resulted in about 361,000 units being replaced, of which 141,000 units were given incentives through the program.

The program is estimated to have avoided approximately 41 MW of peak demand in every year of the program. The program costs around $19.9 million over 8 years, or approximately $490/kW of peak-load reduction.[17]

In conclusion, these three programs document the potential for NYSERDA demand programs to cost-effectively reduce peak loads.

Estimating the Potential for Demand Reduction

The committee estimated the potential for demand-

---

[14]For more on these programs, see the useful tables in "New York Energy $mart Program Cost-Effectiveness Assessment" (NYSERDA, 2004b, p. 2-3).

[15]An updated program evaluation report (Heschong Mahone Group, 2005) evaluated the Peak Load Reduction and Enabling Technologies Programs together. It estimates peak reductions of 178 MW (p. 25), costs of $28.8 million (Table 3-9, p. 24), for a cost per peak reduction of $163/kW.

[16]See NYSERDA (2004b, p. 34).

[17]An updated program evaluation report (Heschong Mahone Group, 2005) estimates peak reductions of 19.7 MW (Table 3-1, p. 16), costs of $18.4 million (Table 1-3, p. 4), for a cost per peak reduction for $934/kW.

response programs to reduce peak demand in the Indian Point service area, as shown in Table 2-5. Details of the estimation are provided in Appendix G-3, "Estimating Demand-Response Potential."

In total, energy-efficiency and demand-response programs in Zones I, J, and K are estimated to be able to deliver peak-demand reductions of 150 MW in 2007, rising to 650 MW in 2010, and 875 MW in 2015 (see Tables 2-4 and 2-5).

The Potential for Expanded Combined Heat and Power

Many studies have assessed the potential for combined heat and power in New York State, with some looking more specifically at opportunities within the Consolidated Edison service territory and/or the relevant New York Control Area load zones in the vicinity of Indian Point.

A 2002 study in New York State (NYSERDA, 2002) noted that there are approximately 5,000 MW of CHP already installed in the state; it assessed the "technical potential" for additional CHP, that is, "the remaining market size constrained only by technological limits." Technical potential does not consider other factors such as capital availability, natural gas availability, and variations in consumption within customer application and size class. The report looked only at CHP, not at other DG technologies that do not involve heat production. It identifies nearly 8,500 MW of technical potential for new CHP in New York at 26,000 sites. Close to 74 percent of remaining capacity is below 5 MW and is primarily at commercial and institutional facilities.

The largest proportion of this capacity is in the ConEd service territory. NYSERDA (2002) identified almost 3,000 MW of technical potential among its customers, the largest opportunities being office buildings, hotels and motels, apartments, schools, and colleges and universities. The report also identified about 300 MW of CHP technical potential among ConEd industrial customers, the largest opportunities being chemical and food plants and textile and paper manufacturers.

The NYSERDA (2002) study stressed that the actual market penetration of CHP will depend on several factors, including the economic advantage of CHP over separately purchased fuel and power, the sites with economic potential, and the speed with which the market can ramp up in the development of new projects. The study developed base case and accelerated case models for CHP market penetration; the models differed in terms of assumptions about power costs, standby rates, technology advances, CHP policy changes including tax incentives, and customer awareness and adoption rates. In the base case, an additional 764 MW of CHP is projected to be installed in New York State by 2012. Nearly 70 percent of this capacity (or 535 MW) is projected to be in the downstate region that includes Indian Point. In the accelerated case, cumulative market penetration reaches nearly 2,200 MW statewide. About 60 percent (1,320 MW) of the penetration is projected in the downstate region in 2012.

Using a trajectory of market expansion for CHP similar to that for energy-efficiency and demand-response programs, the base case estimate of 535 MW in 2012 could be phased in to the marketplace as estimated by the committee and presented in Table 2-6.

The Potential for Expanded Distributed Photovoltaics

Photovoltaics can provide high-value peak-time power in a distributed fashion and with minimal environmental emissions. Thus, PV could contribute significantly to grid stability, reliability, and security (Perez et al., 2004). Rapidly declining PV costs could make this technology a significant contender for replacement power within the time frame of this study even though PV is an intermittent source of electricity. Throughout the 2006-2015 period, installations would have to be subsidized, but the end result could be an important new energy source with many desirable attributes and a thriving industry.

TABLE 2-5 Committee Estimation of Potential Peak Reduction from Demand-Response Programs in New York Control Area Zones I, J, and K, Selected Years Between 2007 and 2015 (MW)

	Reductions in Year				
	2007	2008	2010	2013	2015
Demand-response programs	50	100	200	275	300

NOTE: Zone I, southern part of Westchester County; Zone J, New York City; Zone K, Long Island outside of New York City. Details of the estimation are provided in Appendix G-3, "Estimating Demand-Response Potential."

TABLE 2-6 Committee Estimation of Potential Peak Reduction from Combined Heat and Power in New York Control Area Zones I, J, and K, Selected Years Between 2007 and 2015 (MW)

	Reductions in Year				
	2007	2008	2010	2013	2015
Combined heat and power	100	200	450	550	600

NOTE: Zone I, southern part of Westchester County; Zone J, New York City; Zone K, Long Island outside of New York City. Details of the estimation are provided in Appendix G-3, "Estimating Demand-Response Potential."

SOURCE: Derived from NYSERDA (2002).

Unlike the options discussed above, projections of PV installations on the scale envisioned here cannot be based on current prices or U.S. programs and progress. Rather, the accelerated PV-deployment scenario described here is modeled on the Japanese program that provided a declining subsidy to residential PV systems over the past decade. Residential PV installations expanded in Japan from roughly 2 MW in 1994 to 800 MW in 2004 (Ikki, 2005). Results are presented in Table 2-7; the analysis is in Appendix D-7, "Distributed Photovoltaics to Offset Demand for Electricity," and Appendix G-4, "Estimating Photovoltaics for Demand Reduction." (The analysis of PV potential is based on solar insolation data from the National Solar Radiation Data Base of the U.S. Department of Energy's National Renewable Energy Laboratory [NREL]. This database has data from seven sites in New York State, including one site in New York City.) It might also be noted that, in January 2006, California announced a solar initiative with a goal of 3,000 MW of photovoltaics by 2017 (California PUC, 2006).

Summary

Additional cost-effective demand-side investments in energy efficiency, demand response, and combined heat and power facilities can significantly offset peak demand, as presented in Tables 2-4 though 2-6. These new initiatives (beyond those currently anticipated) could reduce peak demand by 1 GW or more by 2010 and 1.5 GW by 2015. If the cost of distributed photovoltaics can be brought to near-competitive levels over the next decade (see Table 2-7), demand-side measures could contribute 1.7 GW by 2015, thus approaching the capacity of Indian Point (about 2 GW).

The effectiveness of demand-side options in downstate New York, to date, has been variable owing to numerous obstacles to deployment, and forecasted program performance is always uncertain. However, there is a growing body of evidence from New York (through NYSERDA), California, and other states and communities that demand-side options can be implemented swiftly and cost-effectively. Conclusions for each of the four demand-side opportunities are summarized in Figure 2-4.

Energy efficiency programs offer significant potential for peak-demand reduction. Based on prior assessments of hundreds of energy-efficiency measures for residential and commercial buildings, it is estimated that 100 MW of additional peak reduction could be achieved in 2007 if new and expanded programs were to begin in January 2006. This economic and programmable potential is assumed to grow to 450 MW in 2010 and to reach 575 MW by 2015 (Table 2-4).

The estimated potential for demand-response programs to reduce peak demand in the Indian Point service territory is based on the experience to date with three NYSERDA programs that avoided the need for 715 MW of peak demand in the state of New York in 2004. Evaluations of the recent performance of these programs suggest that they offer a highly cost-effective mechanism for reducing peak demand. Assuming that a doubling of program budgets could expand the demand reduction by 50 percent, the committee estimates that the Indian Point service territory has the potential for

TABLE 2-7 Committee Estimation of Potential Peak Reduction from Photovoltaics in New York Control Area Zones I, J, and K, Selected Years Between 2007 and 2015

	Achieved in Year				
	2007	2008	2010	2013	2015
Installed system cost ($/W)	7.36	7.02	6.34	5.40	4.80
Subsidy rate (%)	47	44	38	27	19 (declining to 0 in 2019)
Annual subsidy (million $)	29	36	56	74	72 (declining to 0 in 2019)
Annual installations (MW)	8.4	11.8	23.0	50.4	78.8
Cumulative installations (MW)	18.6	30.4	69.9	192.9	334.7
Reduction in peak demand (MW)	14	23	52	144	250

NOTE: Zone I, southern part of Westchester County; Zone J, New York City; Zone K, Long Island outside of New York City. Details of the estimation are provided in Appendix D-7, "Distributed Photovoltaics to Offset Demand for Electricity," and Appendix G-4, "Estimating Photovoltaics for Demand Reduction."

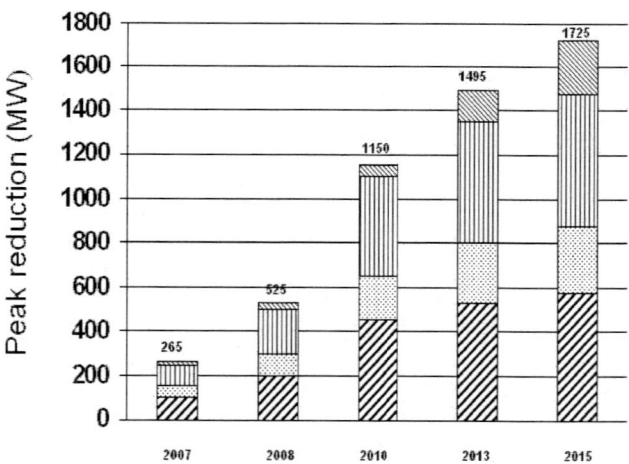

FIGURE 2-4 Phased-in programmable potential for expanded demand-side options in the Indian Point service territory (in megawatts of peak reduction). Heavily hatched bars, energy-efficiency programs; dotted bars, demand-response programs; vertically striped bars, distributed heat and power; and lightly hatched bars, distributed photovoltaics.

expanded summer peak reductions of approximately 200 MW in 2010 and 300 MW in 2015 (Table 2-5).

The actual market penetration of combined heat and power will depend on several factors including fuel prices, standby rates, and the speed with which the market can ramp up its production and services. Under the assumption of accelerated deployment policies, the phase-in programmable potential for expanded CHP is estimated to grow from 100 MW in 2007 to 450 MW in 2010 and 600 MW in 2015 (Table 2-6).

Under an aggressive deployment scenario, it is estimated that 70 MW of distributed photovoltaics could be installed in the Indian Point service territory by 2010, and 335 MW by 2015 (Table 2-7). Realizing this accelerated scenario would require reductions in the cost of PV systems and a long-term commitment to expanding New York's existing PV programs. Such an initiative could establish a self-sustaining PV market in New York, resulting in the continued growth in PV distributed power well beyond the time horizon of this study.

It should be noted that the discussion in this chapter has been relevant to the summer peak only. The New York Control Area also has a winter peak that is about 80 percent of the summer peak. Some of the efficiency measures (e.g., air conditioners) discussed here will not apply in the winter, and PV will contribute little or nothing to the winter peak. The committee did not have the time or resources to examine the winter peak, but this analysis should be performed before it can be fully concluded that demand-side measures would play a large role in replacing the electric power from Indian Point. This analysis also should include a full assessment of the availability of natural gas to enable expanded CHP use in winter (curtailments of gas deliveries to electric generators already occur in the heating season) and the somewhat higher efficiency of many generators and transmission lines in cold weather.

Impediments to Demand-Side Programs

If demand-side programs are so cost-effective, why are they not in more widespread use? If individuals or businesses can make money from energy efficiency, why don't they all just do so? If electricity providers can reduce demand more cheaply than they can deliver new energy supplies, why isn't energy efficiency a larger part of their services? These questions can be answered in large part by describing the range of obstacles that prevent the full exploitation of energy efficiency, including misplaced incentives, distortions of fiscal and regulatory policies, electricity pricing policies, insufficient and incorrect information, and others as discussed below. These are the targets that policies would have to address if demand-side options are to play their full role.

As suggested in that long list, the impediments to energy efficiency are numerous and variable. They depend on the characteristics of a region, the technology, and the supply infrastructure. At the outset, misplaced incentives inhibit energy-efficient investments whenever an "intermediary" has the authority to act on behalf of a consumer, but does not fully reflect the consumer's best interests. The landlord-tenant relationship is a classic example of misplaced incentives. Decisions about the energy features of a building (e.g., whether to install high-efficiency windows and lighting) are often made by people who will not be responsible for the energy bills. For example, landlords often buy the air conditioning equipment and major appliances, while the tenant pays the electricity bill. As a result, the landlord is not generally rewarded for investing in energy efficiency. Conversely, when the landlord pays the utility bills, the tenants are typically not motivated to use energy wisely. As a result, tenants have no incentive to install efficient measures benefiting the landlord, and the landlord has little incentive to invest in measures that benefit the tenant (Ottinger and Williams, 2002). About 90 percent of all households in multifamily buildings are renters, which makes misplaced incentives a major obstacle to energy efficiency in urban housing markets such as New York City.

Distortionary fiscal and regulatory policies can also restrain the use of efficient energy technologies. A range of these obstacles was recently identified in an analysis of projects aimed at installing distributed generation, which is modular electric power located close to the energy consumer; it includes photovoltaics, diesel generators, gas turbines, and fuel cells. Regulatory barriers to these new technologies include state-to-state variations in environmental permitting requirements that result in significant burdens to project developers. Utilities also set high uplift charges (a fee that taxes the amount of revenue gained from selling electricity) and demand fees (a charge that penalizes customers for displacing demand from utilities) that discourage the use of distributed power systems (Allen, 2002). A recent study by the NREL found a variety of "extraneous" charges associated with the use of dispersed renewable technologies (Alderfer and Starrs, 2000). The senior editor of *Public Utilities Fortnightly* described such charges as "a major obstacle to the development of a competitive electricity market" (Stavros, 1999, p. 37).

Electricity pricing policies can also prevent markets from operating efficiently and subdue incentives for energy efficiency. The price of electricity in most retail markets today is not based on time of use. It therefore does not reflect the real-time costs of electricity production, which can vary by a factor of ten within a single day. Because most customers buy electricity as they always have—under time-invariant prices that are set months or years ahead of actual use—consumers are not responsive to the price volatility of wholesale electricity. Time-of-use pricing would encourage customers to use energy more efficiently during high-price periods. These market failures can be exacerbated by competitive wholesale markets, since generators have no incentive to promote efficiency or load management because they profit handsomely from high peak prices. Under current rate designs, LSEs also profit from throughput, finding their profits mitigated by energy efficiency programs. In this way, current market structures "actually block price signals from reaching service providers" (Cowart, 2001, p. vii).

In sum, because of these market barriers, neither electricity generators, transmission companies, nor consumers see the real value of efficiency. Without better price signals, it is challenging for the providers of energy-efficient products and services to transform consumer markets; as a result, incentives such as rebates and tax credits for improved end-use technologies are needed above and beyond those that already exist.

Furthermore, insufficient and incorrect information can also be a major obstacle to energy efficiency. Reliable information about product price and quality allows firms to identify the least costly means of production and gives consumers the option of selecting goods and services that best suit their needs. Yet information about energy-efficient options is often incomplete, unavailable, expensive, and difficult to obtain. With such information deficiencies, investments in energy efficiency are hindered. It is difficult to learn about the performance and costs of energy-efficient technologies and practices because the benefits are often not directly observable. For example, residential consumers get a monthly electricity bill that provides no breakdown of individual end uses, making it difficult to assess the benefits of efficient appliances, televisions, and other products. The complexity of design, construction, and operation of commercial buildings makes it difficult to characterize the extent to which a particular building is energy efficient.

While there are tools such as ENERGY STAR branding, studies have shown that many consumers do not understand them. Further compounding the problem of measuring gains from efficiency concerns the notion of "take-back." When a device has a gain in energy efficiency, consumers have additional resources to spend or save. Some of these resources may be spent on additional energy-consuming activities, which means that the full potential for energy savings does not materialize. Blumstein (1993, p. 970) noted "that low-income programs have a higher than average 'take-back' effect (the participants take back some of the energy saved by taking other actions to increase their comfort)." Based on a recent review of a wide range of markets (Geller and Attali, 2005, Table 1), the take-back, or rebound, effect would appear to be relatively small, generally ranging from 10 to 20 percent.

Decision-making complexities are another source of imperfect information that can confound consumers and inhibit "rational" decision making. Even while recognizing the importance of life-cycle calculations, consumers often fall back to simpler first-cost rules of thumb. While some energy-efficient products can compete on a first-cost basis, many of them cannot. Properly trading off energy savings versus higher purchase prices involves comparing the time-discounted value of the energy savings with the present cost of the equipment—a calculation that can be difficult for purchasers to understand and compute. This is one of the reasons builders generally minimize first costs, believing (probably correctly) that the higher cost of more efficient equipment will not be capitalized into a higher resale value for the building. Moreover, the decentralized nature of the construction industry—home to more than 100,000 builders in the United States—usually means that those engaged in building design and construction have little interaction with one another. The result is lack of information awareness among builders, consumers, and specialists in the building process (Alliance to Save Energy, 2005; Loper et al., 2005). The complexity of the building market is accompanied by confusing and uncoordinated institutional arrangements, with different government agencies sometimes in charge of regulating, implementing, and enforcing the same statute. For example, 18 states have adopted the International Energy Conservation Code of 2003, while 9 states have energy codes that are more than a decade old or follow no energy code at all.

Energy efficiency is not a major concern for most consumers because energy costs are not high relative to the cost of many other goods and services. In addition, the negative externalities associated with the U.S. energy system are not well understood by the public. The result is that the public places a low priority on energy issues and energy-efficiency opportunities, which in turn reduces producers' interest in providing energy-efficient products. In most cases, energy is a small part of the cost of owning and operating a building or a factory. Of course, there are exceptions. For low-income families, the cost of utilities to heat, cool, and provide other energy services in their homes can be a very significant part of their income—averaging 15 percent compared with 4 percent for the typical U.S. citizen. For energy-intensive industries such as aluminum and steel, energy can represent 10 to 25 percent of their production costs. Many companies in these more energy-intensive firms have decided to incorporate energy management as a key corporate strategy.

Since energy costs are typically small on an individual basis, it is easy (and rational) for consumers to ignore them in the face of information-gathering and transaction costs (Harrington and Murray, 2003, p. 3). However, the potential energy savings can be important when summed across all consumers. A little work to influence the source of mass-produced products can pay off in significant efficiency improvements and emissions reductions that rapidly propagate through the economy owing to falling production costs as market shares increase.

Energy prices, as a component of the profitability of an investment, are also subject to large fluctuations. The uncertainty about future energy prices, especially in the short term, seems to be an important barrier. Such uncertainties often lead to higher perceived risks and therefore to more stringent investment criteria and a higher hurdle rate. An important reason for high hurdle rates is capital availability. Capital rationing is often used within firms as an allocation means for investments, leading to hurdle rates that are much higher than the cost of capital, especially for small projects.

Lack of availability of energy-efficient technologies is also often a problem. For example, the purchase of heat-pump water heaters and ground-coupled heat pumps has been handicapped by limited access to equipment suppliers, installers, and repair technicians (Brown et al., 1991; Optimal Energy and the State Grid Corporation DSM Instruction Center, 2005). The problem of access is exacerbated in the case of heating equipment and appliances; because they are often bought on an emergency basis, choices are limited to available stock. Retrofitting can also be expensive, time consuming, and intrusive for home owners and commercial enterprises, especially for businesses that cannot afford the "downtime" needed for installation. Building stock also turns over very slowly, suggesting that inefficient structures remain in use for decades (Ferguson and White, 2003, pp. 15-16).

Finally, managerial and commercial attitudes impede the use of energy-efficient technologies. In the manufacturing sector, energy-efficiency investments are hindered by a preference for investments that increase output compared with investments that reduce operating costs (Hirst and Brown, 1990; Alliance to Save Energy, 1983; Sassone and Martucci, 1984). Similarly, electric utilities believe that they possess the duty and obligation to serve customers' needs. Electric utility regulations have been built on ancient common law duty, known as the "duty to serve" the customer, applied to public utilities such as ferries, flour mills, and railroads. In the words of James Rossi, professor of law at Florida State University, "In the public utility context the duty to serve requires service where it is not ordinarily considered profitable." As one utility executive exclaimed in a recent editorial, "We can't hide behind restructuring and deregulation. Even with unbundled generation, the obligation to serve the load remains" (Lovins et al., 2002, p. 88). Thus, the belief among utility managers and policy makers persists that they need only provide the energy that the customer requires, rather than reforming their customers' consumption requirements through energy-efficiency measures.

Collectively, these social, economic, and cultural impediments greatly inhibit the use of demand-side options. Aggressive policy measures are required to overcome them.

REFERENCES

Alderfer, R. Brent, and Thomas J. Starrs. 2000. *Making Connections: Case Studies of Interconnection Barriers and Their Impact on Distributed Power Projects.* National Renewable Energy Laboratory (NREL)/SR-200-28053. Golden, Colo.: NREL.

Allen, Anthony. 2002. "The Legal Impediments to Distributed Generation." *Energy Law Journal* 23: 505-523.

Alliance to Save Energy. 1983. *Industrial Investment in Energy Efficiency: Opportunities, Management Practices, and Tax Incentives.* Washington, D.C.: Alliance to Save Energy.

—. 2005. *Blueprint for Energy-Efficiency Acceleration Strategies for Buildings in the Western Hemisphere.* Washington, D.C.: Alliance to Save Energy, June 14.

Blumstein, Carl. 1993. "The Cost of Energy Efficiency." *Science* 261(5124, August 20): 969-971.

Brown, M.A., L.G. Berry, and R. Goel. 1991. "Guidelines for Successfully Transferring Government-Sponsored Innovations." *Research Policy* 20(2): 121-143.

Brown, Marilyn A., Frank Southworth, and Therese K. Stovall. 2005. "Towards a Climate-Friendly Built Environment." Pew Center on Global Climate Change, June. Available at http://www.pewclimate.org/. Accessed April 21, 2006.

California PUC (California Public Utility Commission). 2006. *California Solar Incentive Program.* Available at http://www.cpuc.ca.gov/static/energy/solar/. Accessed April 21, 2006.

Cowart, Richard. 2001. *Efficient Reliability: The Critical Role of Demand-Side Resources in Power Systems and Markets.* Report to the National Association of Regulatory Utility Commissioners. Regulatory Assistance Project. June.

Energy Information Administration. 2006. *Annual Energy Outlook 2006.* Available at http://www.eia.doe.gov/oiaf/aeo/index.html. Accessed April 21, 2006.

Ferguson, Richard, and V. John White. 2003. "Risky Diet: 2003." In *Natural Gas: The Next Energy Crisis.* Los Angeles: Center for Energy Efficiency and Renewable Technologies. September.

Geller, H., and S. Attali. 2005. "The Experience with Energy Efficiency Policies: Learning from the Critics." International Energy Agency. Available at www.iea.org/textbase/papers/2005/efficiency_policies.pdf. Accessed April 21, 2006.

Gillingham, K., R. Newell, and K. Palmer. 2004. *Retrospective Review of Demand-Side Energy Efficiency Policies.* Washington, D.C.: National Commission on Energy Policy.

Goldman, C., N. Hopper, R. Bharvirkar, B. Neenan, R. Boisvert, P. Cappers, D. Pratt, and K. Butkins. 2005. *Customer Strategies for Responding to Day-Ahead Market Hourly Electricity Pricing.* LBNL-57128. Berkeley, Calif.: Lawrence Berkeley National Laboratory.

Harrington, Cheryl, and Catherine Murray. 2003. "Who Should Deliver Ratepayer Funded Energy Efficiency?" Regulatory Assistance Project Survey and Discussion Paper. Montpelier, Vermont. May.

Heschong Mahone Group. 2005. "New York Energy Smart Program Cost-Effectiveness Assessment." Prepared for New York State Energy Research and Development Authority. June.

Hirst, E., and M.A. Brown. 1990. "Closing the Efficiency Gap: Barriers to the Efficient Use of Energy." *Resources, Conservation and Recycling* 3: 267-281.

Ikki, Osamu. 2005. *PV Activities in Japan.* Tokyo, Japan: RTS Corporation. May.

Kirby, Brendan, Chuck Goldman, Grayson Hefner, and Michael Kintner-Meyer. 2005. *Load Participation in Reserves Markets: Experiences in U.S. and Internationally.* October.

Komanoff, Charles. 2002. "Securing Power Through Energy Conservation and Efficiency in New York: Profiting from California's Experience." Pp. 1-22 in *Report for the Pace Law School Energy Project and the Natural Resources Defense Council.* May.

Loper, J., L. Ungar, D. Weitz, and H. Misuriello. 2005. *Building on Success: Policies to Reduce Energy Waste in Buildings.* Report to the Alliance to Save Energy. Washington, D.C. July.

Lovins, Amory, et al. 2002. *Small Is Profitable: The Hidden Benefits of Making Electrical Resources the Right Size.* Snowmass, Colo.: Rocky Mountain Institute.

Margolis, Robert M., and Frances Wood. 2004. "The Role for Solar in the Long-Term Outlook of Electric Power Generation in the U.S." Paper presented at the IAEE North American Conference in Washington, D.C., July.

National Energy Efficiency Best Practices Study. 2004. *Best Practices Benchmarking for Energy Efficiency Programs.* Crosscutting Best Practices and Project Summary, December.

NYISO (New York Independent System Operator). 2005a. *Comprehensive Reliability Planning Process Supporting Document and Appendices for the Reliability Needs Assessment.* Albany, N.Y. December 21.

—. 2005b. *Comprehensive Reliability Planning Process and Draft Reliability Needs Assessment.* Albany, N.Y. September 1.

NYSERDA (New York State Energy Research and Development Authority). 2002. *Combined Heat and Power, Market Potential for New York State.*

—. 2003. *Energy Efficiency and Renewable Energy Resource Development Potential in New York State.* Final Report. August.

—. 2004a. *New York Smart Program Evaluation and Status Report.* Final Report: Executive Summary, May.

—. 2004b. *New York Energy $mart Program Cost-Effectiveness Assessment.* Submitted by Heschong Mahone Group. December.

—. 2005a. *2004 Annual Report.*

—. 2005b. *Financial Statements.* March 31. Available at www.nyserda.org/publications/financials05.pdf. Accessed April 21, 2006.

Optimal Energy and the State Grid Corporation DSM Instruction Center. 2005. "DSM Strategic Plan for Jiangsu Province." Final Draft Report, February 11.

Ottinger, Richard L., and Rebecca Williams. 2002. "2002 Energy Law Symposium: Renewable Energy Sources for Development." *Environmental Law* 32: 331-362.

Perez, Richard, et al. 2004. "Solar Energy Security." *REFocus* (July/August): 24-29.

Plunkett, John, and Ashok Gupta. 2004. "State of New York Public Service Commission: Proceeding on the Motion of the Commission as to the Rates, Charges, Rules and Regulations of Consolidated Edison Company of New York, Inc. for Electric Service." December 15.

Rufo, Michael, and Fred Coito. 2002. "California's Secret Energy Surplus: The Potential for Energy Efficiency." San Francisco: The Energy Foundation. September 23. Available at http://www.ef.org/news_reports.cfm?program=viewall&sort=creationdate. Accessed April 21, 2006.

Sassone, P.G., and M.V. Martucci. 1984. "Industrial Energy Conservation: The Reasons Behind the Decisions." *Energy* 9: 427-437.

SEIA (Solar Energy Industries Association). 2004. *Our Solar Power Future: The U.S. Photovoltaic Industry Roadmap Through 2030 and Beyond.* Washington, D.C.: Solar Energy Industries Association.

Silva, Patricio. 2001. "National Energy Policy: Conservation and Energy Efficiency." Hearing Before the Subcommittee on Energy and Air Quality of the House Committee on Energy and Commerce, June 22. Washington, D.C.: Government Printing Office.

Stavros, Richard. 1999. "Distributed Generation: Last Big Battle for State Regulators?" *Public Utilities Fortnightly* 137(October 15): 34-43.

USDOE (U.S. Department of Energy). 2004. *Solar Energy Technologies Program, Multi-year Technical Plan 2003-2007 and Beyond.* Report DOE/GO-102004-1775. Washington, D.C.: Office of Energy Efficiency and Renewable Energy, U.S. Department of Energy.

3

Generation and Transmission Options

When an electric generating plant is retired, it usually is replaced with other generating capacity—perhaps a new generating unit or a new transmission line from an area with surplus power. Either or both reactors at the Indian Point Energy Center could be replaced with these options. However, demand growth projected by the New York Independent System Operator (NYISO) for the New York City area (see Chapter 5) would require considerable additional capacity even without the retirement of Indian Point. That growth can be moderated, as discussed in Chapter 2, but it is likely to be significant. The supply options discussed in this chapter must be adequate to handle growth, retirements of existing capacity, and the potential replacement of Indian Point, if reliability of supply is to be maintained.

This chapter discusses the options for generation, transmission infrastructure, and reactive power in New York. Distributed generation is discussed in Chapter 2 with other end-user options because it generally is not dispatchable by NYISO and is not included in reliability calculations.

EXISTING GENERATING CAPACITY

New York's existing electricity generation is a diverse supply resource, including natural gas, oil, coal, hydroelectric, nuclear, and wind power, as described in Chapter 1. However, much of this generation is far from the large and growing load centers of the New York City area. Western New York (New York Control Area [NYCA] Zones A through E) has surplus of capacity, while New York City (Zone J) is an importer of power, as shown in Table 3-1. The Lower Hudson Valley (Zones G through I) currently has a capacity well above its load, but that will more than disappear if Indian Point is closed. Long Island also must have imported power available to meet its reserve requirement (NYISO, 2005b).

The NYCA, taken as a whole, had approximately 1,300 megawatts (MW) of excess summer resource capability in 2005, representing an excess reserve margin of 3.5 percent.[1]

TABLE 3-1 Approximate (Noncoincident) Summer Peak Load and Capacity in New York State, by Region

Zone	Peak Load (MW)	Capacity (MW)
West (A through E)	8,900	14,430
Upper Hudson Valley (F)	2,180	3,470
Lower Hudson Valley (G through I)	4,490	5,490
New York City (J)	11,150	8,940
Long Island (K, outside of NYC)	5,050	5,180

NOTE: Numbers are approximate and based on the summer of 2004.

SOURCE: NYISO (2005a).

However, the situation by 2008 will be tighter. NYISO expects peak demand to increase by 1,370 MW, and capability may actually decline because of plant retirements. Thus, reserve margins could be lower than the standard requires, even without the retirement of either of the Indian Point reactors.

In addition to the excess capacity in the western section of the state and the Upper Hudson Valley region, some underutilized capacity might be found in the neighboring control areas: the mid-Atlantic counterpart to the NYCA, known as "Pennsylvania Jersey Maryland" [PJM]; Canada; and New England. In the past 5 years, the NYCA imported approximately 10 percent of its energy requirements from PJM and Canada. The annual energy exchange between the NYCA and New England is essentially neutral. It is difficult to determine exactly how much capacity might be found (much of the key information is proprietary) and whether the

---

[1] The NYISO (2005b) report *Comprehensive Reliability Planning Process* lists total capability of 38,772 MW and an expected peak demand of 31,960 MW (demand actually peaked at 32,075 MW in July 2005). The required capability with an 18 percent reserve margin is 37,395 MW. Thus there was an excess capability of 1,327 MW.

transmission capacity (discussed later in this chapter) to deliver it to the New York City area is available. In addition, with demand growing elsewhere and more retirements likely, current excess capacity may not be available in a few years.

Currently, at most only a few hundred megawatts could be imported to the New York City area during peak periods, and demand growth is likely to account for that in a few years (Hinkle et al., 2005; discussed in Chapter 5 of this report). Additional power could be imported during peak periods if the transmission grid was upgraded (and in nonpeak periods even without upgrades).

POTENTIAL NEW GENERATING CAPACITY

Having concluded that the existing generation and transmission system could make little contribution to replacing Indian Point, the Committee on Alternatives to Indian Point for Meeting Energy Needs turned to the question of potential new generation. The committee examined 18 potential alternative generating technologies for possible use in the Lower Hudson Valley/New York City region, including 5 natural-gas-based options, 5 coal-based options, 2 biomass options, 3 wind options, 2 solar options, and 1 advanced nuclear power plant option. Many of these technologies were determined to be unlikely to make a significant contribution to the power needs of the New York Control Area in the time frame of this study. Appendix D-1, "Cost Estimates for Electric Generation Technologies," lists all of the technologies considered with their key cost elements, and Appendix D-2, "Zonal Energy and Seasonal Capacity," presents data for comparisons of zonal energy and seasonal capacity, including the use of supplemental oil with gas turbines.

Technologies Considered

Potential generating technologies include natural-gas-fired units, coal-fired units, biomass-powered units, wind systems, solar-based technologies, and advanced nuclear reactors. Table 3-2 lists the technologies considered and some of their characteristics.

Natural Gas

The use of natural gas as a relatively clean fuel for electric power generation has grown rapidly over the past 20 years as the supplies became more available from various areas of the United States and Canada compared with the period of the mid-1970s. Appendix D-3, "Energy Generated in 2003 from Natural Gas Units in Zones H Through K," shows power generation from natural gas in the New York City area in 2003 and 2004. It also shows that replacing all of Indian Point's power with natural gas would require about a one-third increase in the consumption of gas for electricity.

The technologies that are currently used to convert natural gas to electricity are much more efficient and reliable than earlier versions. The environmental benefits of natural gas relative to other fossil fuels are also a big advantage. Unlike coal, the combustion of natural gas emits no oxides of sulfur, and emissions of nitrogen oxides can be held to standards through stack-gas emission-control systems.

Current supplies of natural gas cannot always accommodate current, let alone increased demand for the product. The owners of gas-fired units in New York State are frequently required to power their gas-fired units with oil products during cold weather periods since the residential sector, with firm delivery service, has priority over the utility sector, which typically has interruptible service tariffs. Generators with backup fuel systems have been providing nearly 20 percent of the electric production derived from the gas turbine facilities in New York State (NYISO, 2005b). For future natural gas turbine facilities to contribute to the electric system during cold weather periods, they should have either backup fuel capability with adequate fuel inventory or firm natural gas pipeline capacity for these periods. Oil tanks could necessitate a larger site footprint, and the combustion of the oil would change the characteristics of the stack-gas emissions, which would have to be addressed. Appendix D-3 lists the oil products used in the overall production of electricity from gas turbines in the New York City area. Peak demand for electricity is higher in the summer than in the winter, and in summer, gas supplies are abundant. Therefore gas supplies are unlikely to affect reliability calculations as discussed in Chapter 5, which focus on the summer peak, but they could well become a constraint during the winter peak. In addition, the increased use of backup oil in the winter raises energy security and environmental issues.

The availability of natural gas in the general area of the Indian Point facility is a key parameter in evaluating alternative generation technologies to replace the two nuclear units. The Algonquin Pipeline system crosses the Hudson River close to the Indian Point power plant on the way to Connecticut. Algonquin's two pipes have a combined capacity of 1.15 billion cubic feet per day (bcf/d), providing natural gas from the Gulf of Mexico into New York and on to New England. New York diverts some 0.12 bcf/d of the gas before it reaches Connecticut. A possibility exists that some of New York's share could be combined with one or more other supplies to assist in generating about 800 MW. The current and future gas supplies would be considered interruptible, since the market environment does not compensate generators for the extra reliability from firm gas supplies or backup fuel supplies.

In addition, a new gas pipeline, the Millennium Pipeline, is currently being installed in New York State. Phase 1 of the project is expected to be complete by November 2006. The line comes from central New York and crosses the Algonquin system near the Ramapo Substation in Rockland County. This line also might supply enough gas for an additional 1,000 MW beyond commitments to customers. The Lovett Power Station site could be served by either line. The

TABLE 3-2 Potential Generating Technologies Considered by the Committee for Replacing Indian Point

Type of Plant	Assumed Capacity (MW)	Relative Potential by 2015[a]	Electricity Cost (¢/kWh)[b]	Output at Peak Demand[c]	Additional Considerations[d]
Natural gas					
Conventional gas combined cycle	250	Large	4.4	High	F, C
Advanced gas combined cycle	400	Large	4.1	High	F, C
Advanced combined cycle with carbon sequestration	400	Small	6.4	High	F, R, D
Conventional combustion turbine (simple cycle)	160	Large	5.8	High	F, C
Advanced combustion turbine (simple cycle)	230	Large	5.3	High	F, C
Coal					
Pulverized coal	600	Large	3.7	High	T, CC
Pulverized coal supercritical	500	Large	3.8	High	T, CC
Integrated coal gasification combined cycle (IGCC)	550	Large	3.7	High	T, D, CC
IGCC with carbon sequestration	380	Small	6.0	High	T, R, D
Fluidized-bed coal	500	Large	4.7	High	T, CC
Renewable energy					
Biomass	80	Small	7.2	High	
Municipal solid waste landfill gas	30	Small	3.5	High	P
Wind					
Large	100	Moderate	5.7	Low	P
Medium	50	Small	6.0	Low	P
Small	10	Small	9.9	Low	
Solar photovoltaics	5	Small[e]	25.0	Moderate	
Solar thermal	100	Small	30.0	Moderate	
Advanced nuclear	1,000	Small	4.2	High	T, P

[a]"Large": the total contribution could be more than 500 MW. "Small": the total is likely to be less than 100 MW. Rated on the basis of readiness of technology, fuel availability, siting difficulties, permitting time, and other factors.

[b]Costs are from Appendix D-1 and are representative for the nation, not the region, which is higher.

[c]"High": virtually all of the maximum capacity can be expected to be available during peak demand. "Moderate": at least half the maximum capacity is likely to be available during peak demand. "Low": it cannot be counted on.

[d]F: additional fuel supply needed; R: research needed; D: demonstration needed; T: additional transmission needed; P: public acceptance questions; CC: high carbon dioxide emissions (>1 lb CO_2/kWh); C: moderate CO_2 emissions (<1 lb CO_2/kWh); no C means little or no CO_2 emissions.

[e]PV may make a significant contribution as a demand-reduction technology, as discussed in Chapter 2.

SOURCE: See Appendix D-1.

three coal-fired units (totaling 431 MW) at the site—on the west side of the Hudson River just across from and south of the Indian Point site—are scheduled to be shut down by 2008, so that site might be available for new gas-fired turbines. Thus, there is likely to be enough gas to supply a significant amount of new capacity at Lovett Station or elsewhere in the area. In addition, other pipelines have been proposed, as shown in Appendix D-4, "Proposed Pipeline Projects in the Northeast." However, two other factors must be considered: namely, the price of gas and other growing demands for the gas (also discussed in Chapter 5).

Current prices for natural gas have been high since the two hurricanes in 2005 damaged some of the infrastructure in the Gulf of Mexico (DOE/EIA, 2005). Also, the overall supply to the state does not appear likely to be increased after the Millennium Pipeline is completed, for the foreseeable future. If so, the New York City area may not be able to continue increasing its use of natural gas for the near term. Furthermore, the longer-term gas supply picture is not encouraging unless resources such as liquefied natural gas (LNG) imports are increased, and LNG imports are uncertain with respect to timing, volumes, and locations for terminal facilities. Investors will have little incentive to build greater pipeline capacity should the supply return only to pre-storm levels in the Gulf region.

Data suggest that gas production from western Canada is declining. Diversions to other users may further limit deliveries to New York. Gas production levels in eastern Canada have experienced poor performance to date, although some gas may become available from Canadian Grand Banks fields. Overall, imports from Canada are not likely to increase significantly unless LNG is routed through Canada. It should be noted that natural gas exploration has increased in the areas south of the Finger Lakes in New York State, and gas production is at record levels for that area (40 bcf per year, or enough for about 800 MW of power generation).

Although it seems as if sufficient gas might be available to replace Indian Point generating capacity, in fact all of the excess may well be committed some time before the plants are shut down. Electricity demand is growing in the New York City area, and several other plants are scheduled to be retired and must be replaced. All new generating capacity currently being built in New York State, over 2,000 MW, is gas-fired. As discussed in Chapter 5, as much as 1,600 MW could be needed by 2010 to meet reliability requirements even without closing Indian Point. Almost all of the generating capacity in the planning stage that could be brought online by 2010 also is gas-fired (883 out of a total of 1,033 MW).

Advanced natural-gas combined-cycle turbine generation facilities can provide reliable and environmentally attractive electric production service to the New York City region, but the production costs are essentially driven by the price and availability of the natural gas obtained from distant sources. At current prices, fuel costs alone are about 4 cents per kilowatt-hour (¢/kWh) in combined-cycle plants and 6 ¢/kWh in simple-cycle plants. In comparison, coal and nuclear plants have fuel costs of only 1 to 2 ¢/kWh, although their operating and capital costs are higher than for gas-fired plants.[2] Table 3-2 shows estimates of the total costs of electricity for all the options considered by the committee. The breakdown by fuel operations and capital are in Appendix D-1, "Cost Estimates for Electric Generation Technologies."

One possibility would be to replace older, simple-cycle gas turbines with modern combined-cycle plants. This switch, called repowering, can result in 50 percent more power from the same supply of natural gas. In New York City, the East River plant is being repowered, and two units at Astoria are expected to be repowered. Other plants could also be considered.

Coal

Coal-based power production provides approximately 14 percent of the electric energy used in New York State, versus some 50 percent for the nation as a whole. No coal-powered facilities are located in Zones H, I, J, or K, but there are two small coal-fired units (at Lovett Station) in Zone G. The major coal-based electric generating facilities are located in western sections of New York State. The amount of coal-based electricity produced in the state decreased by 1 percent between 2004 and 2005. The closing of the Lovett Station coal-burning generators will reduce this even more.

Coal plants require larger sites than do natural gas plants, in order to accommodate the storage of a 30-day supply of coal, associated ash-management systems, and defined areas to accommodate storm-water-management programs. Coal plants, therefore, are located in areas where property values are relatively low. Land values in the Lower Hudson Valley and New York City areas are among the highest in the nation.

Environmental considerations such as stack-gas emissions, noise from unit trains bringing coal and removing ash, and cooling water requirements all contribute to major siting challenges when using any coal-based generation technology in major urban areas. Coal-based technologies that were considered and evaluated with respect to operating costs are discussed in Appendix D-5, "Coal Technologies." Coal-based power plant technologies that could produce power for the New York City region would be located at some distance from the region, requiring long transmission lines. Therefore, the cost of the power would include transmission costs as well as production costs. In addition, some air quality issues could arise, depending on the location of the associated site.

Coal plants also emit more carbon dioxide per kilowatt-hour produced. Technologies are being developed to capture and sequester the carbon dioxide, but that process will add significantly to the cost of the electricity. Appendix D-5 discusses the technology (integrated gasification, combined cycle—IGCC—that will be most appropriate for capture of carbon dioxide).

A new coal plant built upstate from the New York City area might be the lowest-cost replacement for Indian Point, even with a new transmission line. Thus it should be included in the list of options. However, the committee believes that it is unlikely for a coal facility to be permitted and constructed even in upstate New York by 2015, especially considering the uncertainties over carbon dioxide.

Biomass

Biomass represents a renewable fuel source for power generation. In the New York City area, biomass consists of municipal solid waste, sewage sludge, wood waste, agricultural waste, and other residues. Today there are five waste-to-energy plants in the downstate area, with one in Zone H and four in the Zone K area. The total capacity for these five units is 166 MW, and collectively they produced 1,274 giga-watt-hours (GWh) of power in 2004 of the 52,000 GWh generated in Zones H, I, J, and K. Methane derived from biomass sources can be burned in gas turbines, and biomass in a solid form can be burned directly or gasified. It also can be co-fired in coal-based plants, but as noted above, coal plants are unlikely to be sited in the zones of interest for a variety of reasons.

In the 1980s, there was a move to have a waste energy facility located in each of the five counties of New York City as a measure to assist the city in managing its wastes and to address the need for fuel diversification in the city. The plan was dropped by the New York City government primarily because of strong and widespread public opposition to waste-to-energy plants being located in the city. The principal concerns were air quality and health issues. Municipal solid waste and sewage sludge currently produced in the city are shipped out of state, even though today's technologies are cleaner and might engender less public resistance.

---

[2]Locational-based marginal prices for the NYISO-run wholesale power market are given at https://www.nyiso.com/public/market_data/pricing_data.jsp. Accessed March 2006. As an example, the 4:00 p.m. wholesale clearing price of electricity on January 23, 2006, was 11.9 ¢/kWh in New York City.

Biomass appears unlikely to be a significant new source of electricity for the New York City region. Additional information on the potential of the biomass resources is contained in Appendix D-6, "Generation Technologies—Wind and Biomass."

Wind

Wind energy systems have entered the New York State market with some 100 MW of capacity installed by 2005, and more is expected. The wind facilities are located in the central and northern areas of the state. The New York State Energy Research and Development Authority (NYSERDA) has initiated a wind development program that is installing some 500 MW of new wind capacity as a component of the State's Renewable Portfolio Standard development program. This program mainly provides support to developers after the units are placed into service. The developer has the responsibility to site, license, construct, and place into service its wind facility.

New York State has several excellent wind sites that are being evaluated by developers for near-term application. At this point, few land-based sites are located close to the Indian Point facility that have the desired wind characteristics and available land to install wind turbines that could contribute to the replacement of the generation from the Indian Point plants. A project has been proposed at a site in the ocean off the south shore of Long Island. This project is proceeding, but at a pace slower than originally anticipated, owing to rising costs. Experience with offshore wind projects is limited, and the developers are monitoring projects located elsewhere in the world. The Long Island project and other offshore sites have the resource potential for considerable generation of electric power, but no units have been installed there, and considerable opposition can be anticipated, as has occurred in Massachusetts.

Technically there is sufficient wind resource in New York State to replace the Indian Point units, but resolving site location and permitting issues is key to successfully placing units into service. The greatest challenge for using wind to replace large baseload electric generation units is the intermittent nature of the resource. The availability factor for wind is 30 to 40 percent, compared with about 90 percent for nuclear and coal plants, and the resource is available only when the wind is blowing, not when demand is high. Storage will smooth out the intermittent nature of the resource, but that technology is not yet readily available. The issues associated with expanding the use of wind in the state are discussed in Appendix D-6.

Solar

Solar energy can be used to generate electricity either through the use of solar photovoltaic (PV) systems or through solar thermal power generation technologies. Solar PV electricity is increasingly being used for many applications around the world.

PV use has increased as the price of solar cells and the resultant power costs have decreased and the reliability of the products has risen to a level that is acceptable to consumers for some applications. PV applications are limited by the dependence on the availability of sunlight, but for some applications either that does not matter or else a small amount of battery storage can suffice. The technology promises to grow substantially in the distributed-generation-systems market, as discussed in Chapter 2. PV would require large land areas to collect sufficient energy to contribute to the bulk power markets and is unlikely to be a factor in New York State by 2015, but rooftop-mounted systems supplying directly to the retail market could become significant.

Solar thermal generation involves the use of mirror-like collectors designed to focus sunlight onto metal surfaces, which in turn through various systems can produce a steam product. The steam is then used in a steam turbine to produce electricity. One advantage of the solar thermal concept is that the energy of the Sun can be stored in a liquid material on a clear day and then later extracted to produce steam at night or on cloudy days. Solar thermal generation requires large land areas to house the collectors and very direct sunlight to be economically attractive. The earliest applications of solar thermal technologies will be in the deserts of the southwestern part of the United States. The specific characteristics of the PV technology are discussed in Appendix D-7, "Distributed Photovoltaics to Offset Demand for Electricity."

Advanced Nuclear

Several advanced nuclear technologies are being explored for possible application in the 2015-2020 time frame (EPRI, 2005). The concepts are being supported through programs initiated in part by the recently enacted federal Energy Policy Act of 2005. The Nuclear Regulatory Commission has certified three designs, which could be started shortly after an appropriate site is found and certified. Several consortia of energy companies (including Entergy Corporation) are moving forward on various plans. A site at Oswego, New York, on Lake Ontario, had been considered but is not part of any current plan. That site had strong local support and may be considered in future plans.

Nuclear power could provide New York State with an electric power option that has no carbon dioxide emissions (which contribute to global warming), and no contribution to acid rain or mercury contamination. However, the committee concluded that a new nuclear plant in New York State is unlikely before 2015. One or two of the projects now being planned in other states might be completed by 2015, but most companies are likely to wait in order to see how these plans progress before starting more projects.

Overall Considerations

A variety of supply options could contribute to replacing one or both reactors at the Indian Point Energy Center. As suggested in the previous discussion and in Table 3-2, the committee concludes that advanced natural-gas-fired combined-cycle plants are the generation option capable of making the biggest contribution at the lowest cost by 2015. This position assumes the ability to site such facilities in the Lower Hudson Valley/New York City area, favorable economic and regulatory conditions for investors, sufficient advance notice that the power will be needed, and a long-term fuel supply.

One option that could be considered in the near term is to locate some 2,400 MW of natural-gas-fired combined-cycle plants at the current Lovett Station site, described earlier in this chapter. The site is currently being used for electric production. However, the current operator is just emerging from bankruptcy and may not be in a position to develop any new facilities. If that issue can be resolved, the site could be developed for natural-gas- and/or oil-fired generation. The site has a transmission corridor, with limited transmission currently installed, a developed waterfront, and basic elements of infrastructure. However, environmental impacts would need to be addressed, as would fuel delivery.

The greatest challenge would be to secure sufficient natural gas supplies to satisfy the projected production levels, including very high capacity factors. Two large natural gas lines are located near the Lovett Station site, and more natural gas might be added to the two existing systems from gas wells located in the state. If new sources of gas and new pipelines are required, the issues of gas availability and price must be examined in much greater detail than that allowed by the committee's resources.

Coal-based technologies potentially offer attractive production costs, but the physical requirements of a large plant site in the region of the Indian Point Energy Center, combined with air quality issues, new rail lines to bring in the coal, and related technical challenges limit potential opportunities for investors to promote this fuel source for application in the greater New York City area. If natural gas prices remain high, a coal plant upstate with a new transmission line to the New York City area might be a cost-effective solution.

Both natural gas and coal plants emit carbon dioxide (coal plants emit about twice as much per kilowatt-hour as natural gas plants), which nuclear plants do not. New York is part of the Regional Greenhouse Gas Initiative (RGGI), which proposes to limit emissions of carbon dioxide and other greenhouse gases. Achieving RGGI goals will be more difficult if Indian Point is replaced, as discussed in Chapter 4.

New York State is supporting renewable energy development for power production, including a recently adopted Renewable Portfolio Standard. Nevertheless, renewables are unlikely to provide the Lower Hudson Valley/New York City area with a significant share of the power provided by Indian Point within the time frame of this study.

ELECTRICAL TRANSMISSION

Existing Transmission

Most Americans are generally unaware of the vast electrical transmission network that connects a myriad of power-generating stations to the local power lines servicing their homes and businesses. Electricity is typically generated in large central power stations at 13,800 volts (13.8 kV) then often "stepped up" to 345 kV through power transformers and associated equipment in order to transmit the power efficiently over long distances. These high-voltage transmission lines provide the backbone for the bulk electrical power system throughout the United States. Transmission lines, however, can be designed to be operated at voltages other than 345 kV. Other typical voltages for transmission lines in the United States include 765 kV, 500 kV, 230 kV, 138 kV, 115 kV, and 69 kV. Power system engineers select the optimal voltage for a particular transmission line based on a number of design considerations, including the line's proximity to generation and customer load. In general, however, transmission lines with higher voltages are utilized to interconnect generating plants to the bulk power system.

The bulk power system in New York State is similar to that in many other regions throughout the United States and Canada. According to NYISO, the bulk power system in New York State, the New York Control Area, contains more than 10,000 miles of transmission lines with voltages equal to 115 kV and more. Figure 1-1 in Chapter 1 shows the major transmission facilities in the NYCA with voltages of 230 kV and greater.

The NYCA is electrically connected to neighboring control areas in the northeastern United States and the Canadian provinces of Quebec and Ontario through special high-voltage transmission lines, often referred to as "ties" or "interfaces," such as those shown in Figure 1-1. The total nominal transfer capability between the control areas in the Northeast is less than 5 percent of the total peak load of the region and is declining as a percentage of such load (NYISO, 2005b). This minimal import and export capability over the ties among the Northeast regional control areas means that the NYCA power system places even greater reliance on the internal generation resources located within a particular control region.

Transmission constraints or "bottlenecks" are not just associated with the constrained ties between New York and its neighboring control areas, however. The NYCA has several major transmission bottlenecks within New York State, which significantly affect the free flow of power on its bulk transmission system. In particular, the electrical transmission system around southeastern New York State, including greater metropolitan New York City and Long Island, is severely constrained owing to a lack of adequate transmission capacity into this area. As a result of the limited transfer capability into southeastern New York State, this subregion must place greater reliance on the generating plants located within greater metropolitan New York City and Long Island. As shown in Chapter 5, a new transmission line could deliver a large fraction of the power provided by Indian Point.

Table 3-3 and Figure 1-1 further describe the approximate location of the three major transmission constraints within the NYCA. The Total East Interface constrains power flowing from western New York State, PJM, and Canada into eastern New York State. The Central East Interface is located east of the Total East Interface and serves to further constrain power flowing from the west and central portions of the NYCA. Finally, the Upstate New York-Southeast New York (UPNY-SENY) Interface severely constrains power flowing into southeastern New York State from the rest of New York and from PJM and Canada.

NYISO has segmented the NYCA into 11 distinct zones, as explained in Chapter 1, to accommodate the location of the transmission interfaces and to respect the service territories of the transmission owners. These NYCA zones (see Figure 1-3 in Chapter 1 of this report) function as separate pricing zones under the locational-based marginal pricing (LBMP) wholesale power market operated by NYISO. Given the limited transfer capability shown in Table 3-3 at the transmission interfaces, and the supply-and-demand balance for electricity, the southeastern New York zones (Zones H, I, J, and K) experience the highest average and peak prices within the NYCA. Table 1-1 in Chapter 1 shows the approximate consumer load and associated generating capacity in each NYCA zone. Generating plants in southeastern New York are particularly valuable because they are on the high-demand side of the constraints. The Indian Point generating plant is located in the premium southeastern New York Zone H; hence the consumers in Zones H, I, and J heavily rely on it to meet demand. It is therefore very important to take the bulk transmission system into account when the retirement of Indian Point Units 2 or 3 is considered.

TABLE 3-3 Nominal Transfer Capability Between New York Regions

Transmission Interface	Transfer Capability (MW)
Total East	6,100
Central East	2,850
Upstate New York-Southeast New York	5,100
Cable	
New York City	4,700
Long Island	1,270

SOURCE: New York Independent System Operator.

New Transmission

New transmission capacity, if designed to adequately increase the transfer capabilities among the Total East, Central East, and UPNY-SENY Interfaces, may provide a partial solution to the retirement of Indian Point, including system reliability benefits. Such new transmission capacity would likely come in the form of either an expansion of the existing high-voltage alternating current (HVAC) transmission systems or the addition of new high-voltage direct current (HVDC) transmission facilities.

New AC transmission facilities may include the replacement of conductors on existing transmission facility structures or the installation of new transmission facilities including new tower structures and related components. Such new AC transmission facilities may also require additional right-of-way land resources and potential system outages during construction periods. An expansion of the existing AC transmission system would likely serve to increase system reliability and decrease the marginal cost of electricity in southeastern New York.

New AC transmission facilities may also be coupled with dedicated generation resources to further support New York's "in-city" generation requirements. An illustrative example of such a new AC transmission facility would be the proposed 550-MW Public Service Electric & Gas (PSEG) Cross Hudson Project. That project includes the interconnection of an existing 550 MW natural-gas-fired combined-cycle generating unit located at a New Jersey-based utility, PSEG's Bergen generating plant, with the Consolidated Edison substation at West 49th Street in New York City via underground 345 kV transmission conductors and associated facilities. Combinations of dedicated power-generating resources and interconnection facilities such as the PSEG Cross Hudson Project may offer additional alternatives to adding new generation resources directly into transmission-constrained zones such as Zones H, I, J, and K. However, as useful as this project could be, it is currently inactive and may not be revived.

HVDC transmission projects may also provide partial solutions to the loss of Indian Point Units 2 and/or 3. Such HVDC transmission projects typically require the installation of an AC/DC converter station, HVDC conductors, and a DC/AC converter station. The process entails the conversion of alternating current to direct current (in the AC/DC converter station located near a sending substation), transmission of the power (typically long distances) through high-voltage direct current conductors, and finally the conversion of direct current to alternating current (in the DC/AC converter station) adjacent to the receiving substation. Because an HVDC line is isolated from the regular HVAC grid, it is not subject to the same reliability issues, and the power that it delivers is considered to be equivalent in reliability to that from a plant within the zone of the end point. In particular, New York City and Long Island (Zones J and K), which have requirements for locally produced power (80 and 98 percent, respectively), obtain the same reliability benefit from a dedicated HVDC line as they would from a local power plant. The Neptune transmission line from New Jersey to Long Island will provide reliability benefits as well as cheaper power when it commences operation in 2007.

The addition of a new 1,000 MW HVDC transmission facility between Marcy and Rock Tavern Substations could serve as a suitable alternative to the compensatory action of adding 800 MW of new generation in Zone J. This alternative also serves to increase New York's statewide electric system reliability and could lower total system production costs within the greater Northeast region, including New York State. Further, an additional benefit may include a reduction in imports of electricity from outside the Northeast region owing to the more efficient use of indigenous generation located in upstate New York and PJM (Hinkle et al., 2005).

In summary, it is clear that new transmission projects can play an important role in the ultimate energy and capacity solution relating to the potential loss of power from the Indian Point units. It is likely that a combination of modifications to the existing AC transmission system and the installation of new HVDC transmission projects will provide the best complement to the addition of new generating resources and efficiency programs to solve New York's future electricity needs.

RELIABILITY AND REACTIVE POWER

Reliability

Most of the power interruptions of the typical customer are brief, affecting only a small area, although even very short interruptions that disturb computers and voltage variations that affect voltage-sensitive equipment can be damaging. Many power interruptions are due to local problems, such as an automobile accident knocking down a power distribution pole or a squirrel getting inside a vulnerable piece of equipment in a substation. Outages in distribution systems are outside the scope of this report, which is concerned with the bulk power system.

When the transmission system goes down, perhaps due to severe weather, earthquakes, or multiple equipment failures, entire regions can be blacked out, and recovery can be lengthy. Very large multistate disturbances such as that experienced in August 2003 are rare and involve a combination of many unlikely events. Reliability is measured by the frequency, duration, and magnitude of interruptions and other adverse effects on the electric supply.

The regional reliability councils formed after the 1965 Northeast blackout (New York is in the Northeast Power Coordinating Council) have tried to quantify these disturbances by requiring a measure of reliability based on computing the likelihood that the demand for power cannot be met. Load is modeled as a demand for power that is weather-dependent and varies with the season, the day of the week, and even the hour of the day. The maximum load tends to occur on the hottest summer days. Statistical descriptions of the historical availability of each generator are used to compute the expected number of days in a 10-year period when the load could not be supplied (the loss-of-load expectation, or LOLE). The New York State Reliability Council requires that the number be less than 1 day in 10 years. Changes in the system that would increase the LOLE to more than 1 day in 10 years would not be acceptable.

It is unusual for a blackout to occur simply because a large number of generators were unexpectedly out of service (the 1965, 1977, and 2003 blackouts were much more complicated). Nevertheless, the LOLE is useful in determining how much extra generation a given area requires. Meeting this standard in the NYCA usually means that the available capacity (the total power of all generators able to be scheduled to serve the load) should exceed the peak load by 18 percent.

Because power can be imported from neighboring areas, the reliability and capacity of both the transmission system and the generation equipment must be included in the analysis. The loss of transmission lines to other areas (notably New England, PJM, or Canada) could have serious consequences on a hot summer day. Relief from other control areas is limited, however, as interarea transmission capacity is about 5 percent of peak load and is decreasing with time. A reliable power system has enough excess installed generating capacity so that the load can be supplied even if some generators are out of service for maintenance or because of unexpected problems, and it has a transmission system that is adequate to transport the power from wherever it is generated (inside or outside the control area) to the customers. The mix of generation normally includes some inexpensive baseload generators that tend to run at a constant output around the clock and serve the minimum (base) load, along with units that respond more rapidly to changes in demand and can follow the peak. Nuclear units are operated as baseload units because they usually have the lowest variable operating costs.

An additional reliability concern is the supply of fuel for generators. The adequacy and diversity of fuel constitute an important issue in operating the system and planning new generation. Heavy reliance on a single fuel source or a single pipeline for natural gas could have serious consequences if this supply were interrupted. The competing demand for natural gas for heating in the winter must also be considered as most gas-fired power plants in New York operate on interruptible gas-supply contracts, and therefore most are dual-fuel units that can be switched to oil firing. On an annual basis, however, as noted in Chapter 2, dual-fuel units in New York use natural gas for about 82 percent of their annual generation.

Reactive Power

Major power system disturbances have, in one way or another, involved unstable oscillations of electrical quantities. Dynamic changes in power flows, or in system frequency (departures from 60 hertz), or in voltage reduction are all signs of system instability. Frequency excursions take place when the balance between supply and demand for power is upset. Too much demand produces a lower frequency, and too much supply results in a higher frequency. As the power system came apart in August 2003, there were islands with excess generation and islands with too little generation.

There is another kind of power in alternating current systems, associated with the magnetic fields produced by currents flowing in transmission lines, generators, and motors. This power is called *reactive power* and is measured in vars (for volt-ampere reactive).[3] Reactive power represents energy stored in the magnetic field and later released. Motors such as those in air conditioners and refrigerators also require reactive power to function correctly.

Reactive power also is essential for the smooth operation of the transmission grid. It helps hold the voltage to desired levels. Inadequate reactive power leads to a decrease in the voltage of the system in which the shortage exists. For an interconnected system where active power is exactly in balance, the frequency is constant and the same everywhere, and the system is said to be in synchronous operation. Voltage, however, varies from location to location, depending largely on the reactive power balance. If a given load has a large reactive demand, the voltage will be lower at that point than at others. Low voltage can damage equipment and, if low enough, can cause system instability and a voltage collapse. There have been a few voltage collapses solely because of a shortage of reactive power. It is more common that reactive power problems aggravate active power problems in large power system disturbances, as was the case in the August 2003 event (U.S.-Canada Power System Outage Task Force, 2004).

Active power can be transmitted over great distances, while reactive power problems must be solved locally. Generators themselves are an excellent source of reactive power but at some cost. Increasing the reactive output of a generator results in a decrease in the possible active power output and, if not specifically compensated, a loss of income received for real power output. Capacitors can be a second source of reactive power by storing energy in electrostatic fields rather than electromagnetic fields. Capacitors can be fixed or variable in size. Distributed generators—for example, microturbines and synchronous motors—can also supply reactive power, but these units are outside the control of the system operator and cannot necessarily be counted on when needed.

Indian Point is a large supplier of reactive power to the grid in southeastern New York State, capable of providing about 1,000 megavars of reactive power. If it is shut down, that reactive power must be replaced. Insofar as replacement generation is located upstate or even farther away, it becomes even more important to ensure adequate supplies of reactive power. That could be done by installing capacitors at the Indian Point site or in the general area. Generating vars is not expensive, but it is a critical necessity that must be planned for if Indian Point is to be closed.

NYISO projects that, even with Indian Point operating, voltage constraints due to reactive power deficiencies in the Lower Hudson Valley will lower system reliability to unacceptable levels. Consequently, NYISO has solicited market-based and regulated backstop solutions to correct the reliability deficiency.[4]

REFERENCES

DOE/EIA (Department of Energy/Energy Information Administration). 2005. "Natural Gas Weekly Update." December 22. Available at http://tonto.eia.doe.gov/oog/info/ngw/ngupdate.asp. Accessed December 22, 2005.

EPRI (Electric Power Research Institute). 2005. "Making Billion Dollar Advanced Generation Investments in an Emissions-Limited World." Background paper for the EPRI Summer Seminar, August 8-9, 2005, San Diego, Calif.

Hinkle, G., G. Jordan, and M. Sanford. 2005. "An Assessment of Alternatives to Indian Point for Meeting Energy Needs." Unpublished report for the National Research Council, GE-Energy, Schenectady, N.Y., December 19.

NYISO (New York Independent System Operator). 2005a. *Comprehensive Reliability Planning Process*. October 25.

—. 2005b. *Comprehensive Reliability Planning Process (CRPP), Reliability Needs Assessment, and NYISO Comprehensive Reliability Planning Process, Supporting Document and Appendices for the Draft Reliability Needs Assessment*. December 21.

U.S.-Canada Power System Outage Task Force. 2004. *Final Report on the August 14, 2003 Blackout in the United States and Canada*. April. Available at https://reports.energy.gov. Accessed March 2006.

---

[3] Active power, the familiar type of power that keeps lightbulbs burning, is measured in watts. Consumers pay for active power (1,000 watts used for an hour is a kilowatt-hour) but usually not for reactive power.

[4] See M. Calimano, NYISO solicitation letter to S.V. Lant, R.M. Kessel, E.R. McGrath, and J. McMahon, December 22, 2005.

4

Institutional Considerations and Changing Impacts

The previous two chapters identified the demand- and supply-side options for replacing the generating capacity of the Indian Point Energy Center's two operating nuclear reactors. Putting these options into action in planning and administering the New York Control Area (NYCA) electrical system must be done in the context of economic, social, and institutional impacts as well as with regard to the technological opportunities and constraints. This chapter reviews the most significant general, statewide considerations:[1]

- *Financial underpinnings of the electrical supply system* (that is, how the various organizations that generate, transmit, and distribute power underwrite the necessary investments to ensure reliable service) and how that relates to the evolving institutional structure in New York State; and
- *Environmental and other impacts that affect society.*

REGULATION, FINANCE, AND RELIABILITY

Financial and economic considerations will have a profound effect on the choice of options to replace Indian Point, the reliability of the system, and the costs of substituting generation or transmission options for the Indian Point units. Procedures for maintaining the reliability of the New York State system are discussed mainly in Chapter 5.

The New York State Electricity Market

The impact of the replaced costs of the Indian Point units if they are shut down is dictated by the evolving New York State competitive market and by the socioeconomic background in the state. Indian Point's replacement costs to the customer are virtually impossible to project at present, given the electricity market operation and its evolving status. The reasons are summarized in Boxes 4-1 and 4-2, on the cost of replacing Indian Point: "In Theory" and "In Practice."

This section provides background information on the regulatory and financial environment in New York State and on how this environment shapes the incentives for investing in generation and transmission facilities. It also explains why there are growing concerns about the continued reliability of electricity supply, particularly in New York City. Appendix E, "Paying for Reliability in Deregulated Markets," gives a fuller account of how the regulation of the electric utility industry in New York State has changed and the implications of these changes for reliability.

In response to a number of financial problems, such as the cost of building excess generating capacity in the 1980s, the Federal Energy Regulatory Commission (FERC) supported new legislation in the 1990s to facilitate increased competition in the electric power industry. Competition was introduced initially in the northeastern states and in California, regions that had relatively high prices for electricity under traditional regulation. In 1999, regulators in New York State took the first major step by introducing new markets for electricity (real energy) and ancillary services, such as reserve generating capacity. At the same time, the New York Independent System Operator (NYISO) was established to run these new markets and to control the operation of all power plants in the New York Control Area. Unlike the generation components of the industry, the transmission and distribution components continued to be regulated by the New York Public Service Commission (NYPSC).

Appendix E explains that the current patterns of spot prices in the NYCA have changed and are now much less volatile, with fewer price spikes than when the market was first introduced in 1999. This change in price behavior has made prices more predictable, but at the same time it has reduced the financial earnings of peaking capacity (generating units that are used only to meet relatively short periods of peak demand and therefore have low capacity factors) relative to those of baseload capacity. The consequences of

---

[1] Specific plant and transmission line siting issues, including costs and environmental constraints, are not discussed here, since they vary so widely throughout the state and are considered beyond the scope of the study.

BOX 4-1
The Cost of Replacing Indian Point: In Theory

The cost of replacing Indian Point is substantial because its two operating nuclear reactors, Units 2 and 3, represent 2,000 megawatts (MW) of baseload capacity with relatively low operating costs. In addition, a large capital investment of these units has already been made. To the extent that a replacement strategy includes conventional generating capacity (e.g., using natural gas as a fuel), the incremental cost of building this new capacity will include the capital costs, and in addition, the operating costs will be higher. Under traditional regulation, all of these incremental costs would be passed on directly to customers in New York State. Although someone has to pay for these higher costs, customers may not see major increases in their monthly bills in the new deregulated market in the state. How is this possible? An explanation follows using a simple example of the magnitudes of the costs involved.

Let us assume that the full operating costs of Indian Point are $20 per megawatt-hour (MWh) and that the units operate for a total of 8,000 hours per year. These operating costs would include the nuclear fuel, labor, and capital costs for operations and maintenance (which might require adding a cooling tower in the future), and payments into a sinking fund to cover decommissioning as well as a charge paid to the federal government to cover the cost of disposing of nuclear waste. Since Indian Point has a capacity of 2,000 MW, the total annual cost of operations is $320 million per year (20 x 2,000 x 8,000).

The average wholesale price of electricity in New York Control Area Zone H was $80 per MWh in 2005 (when the price of natural gas was substantially higher than historical levels). Consequently, the annual revenue, if all power had been sold in the wholesale market, would be $1,280 million per year (80 x 2,000 x 8,000) and the annual earnings for Entergy Corporation (the plant's owner) would be $960 million per year (1,280 – 320). The situation is more complicated in reality, because Entergy may have long-term contracts to sell some of the power at prices below the current high level in the wholesale market. Nevertheless, these contracts will have to be renewed periodically, and with high prices for natural gas, Indian Point represents a very valuable source of income for Entergy.

To keep the example simple, let us assume that Indian Point is replaced completely by 2,000 MW of combined-cycle capacity using natural gas as a fuel. The operating cost of these units is $60 per MWh, and the annualized capital cost is $120 per kilowatt per year (kW/year). These units will also operate for 8,000 hours per year, and as a result, the capital cost prorated to the annual amount generated corresponds to $15/MWh (120,000/8,000). The total annual cost of generation is $1,200 million per year ([60 + 15] 2,000 x 8,000), and the incremental cost of replacing Indian Point is $880 million per year (1,200 – 320). That is a very large amount of money, but it could be much lower for a number of valid reasons. For example, reducing load by improving the efficiency of appliances is shown in Chapter 2 of this report to be much more cost-effective than building new generating capacity, and the transmission upgrades discussed in Chapter 3 may allow existing units in other locations to generate more power.

Under traditional regulation, all prudent operating costs and capital costs for generation, transmission, and distribution are aggregated to determine the size of the revenue requirement and the corresponding retail rates charged to customers.[1] In a competitive market for generation, the most expensive unit needed to meet the load sets the wholesale price paid to all units that are generating in the market (prices actually vary from location to location owing to congestion on the transmission lines, but this is not an important issue for this example). When an expensive peaking unit sets the price on a hot summer day, the wholesale price paid to generators is much higher than the operating costs of most units. This "extra" income can be used to cover the capital cost of generation.

In theory, the wholesale price in a competitive market should cover all of the operating and capital costs of generation, but, as explained in this chapter and in Appendix E, "Paying for Reliability in Deregulated Markets," a truly competitive market will not cover the capital cost of a peaking unit unless high prices (scarcity prices) are allowed. However, the total cost of the combined-cycle unit in this example ($75/MWh) is covered by the wholesale price ($80/MWh). Although these results are clearly sensitive to the assumptions made, this specific example shows that it is quite possible in a competitive market to add new generating capacity without increasing the wholesale price. In fact, the simulated market prices in some of the scenarios presented in Chapter 5 are lower when new generating capacity is added. The reason is that the new efficient units displace some generation from existing units that are more expensive to operate, and the more efficient units set the market price more frequently.

Who does pay for the incremental cost of replacing Indian Point in this example, if customers still pay the same wholesale price as before? The main loser in this example is Entergy, because the substantial annual earnings from Indian Point have now been eliminated. Given the many complexities of determining costs, such as the effect of increases in the use of natural gas on the future price of natural gas, it is extremely difficult to measure the true cost to customers of replacing Indian Point. The most important complications about determining this cost are discussed in Box 4-2. The main point of the present example is to show that the current wholesale price of electricity in the New York market may cover a large part of the incremental costs of replacing Indian Point. In a competitive market, the financial consequences for customers are likely to be smaller than the consequences would have been under traditional regulation. There is, however, an important qualification that should be made. The example here and the scenarios presented in Chapter 5 assume that new generating capacity will be built in a timely way before Indian Point is retired. If Indian Point experienced an unscheduled failure and had to be taken off-line in an emergency, the wholesale price would increase substantially. Without Indian Point and without new capacity, more-inefficient units with higher costs would have to be used to meet load. These expensive units would set higher wholesale prices.

---

[1] In fact, traditional regulation did not apply to Indian Point Unit 3, because it was owned by the New York Power Authority, and its power was sold in part outside the regulated market.

> **BOX 4-2**
> **The Cost of Replacing Indian Point: In Practice**
>
> Although the cost of building and operating new electric generating capacity to replace some or all of the 2,000 MW at the Indian Point Energy Center would be substantial, it is very difficult to determine what the overall effect would be on the bills paid by customers. The committee's scenarios, presented in Chapter 5, project the basis for the wholesale market prices in different zones. Generally, these prices are higher than the prices in the base case with Indian Point operating, but in some situations they are lower. The explanation for getting lower wholesale prices is that new efficient capacity displaces some of the old inefficient capacity and sets the market price more often.
>
> The pricing mechanism used in all of the scenarios is based on a uniform-price auction assuming that the market is competitive (i.e., that the offers submitted into the auction by generators are equal to the true production costs, and under this specification, it would be extremely unlikely for the market price ever to be set by the low production cost of Indian Point). Assuming that the market is competitive is a reasonably close representation of how the market is actually performing at this time. Hence, the predicted prices in the scenarios provide a consistent way to determine how wholesale prices would be affected in different situations. Higher wholesale prices would result in higher rates charged to customers unless there was an offsetting reduction in the other costs of generation.
>
> The main complication for determining the total cost of generation in the current market structure is that the wholesale price of electricity is only one of the components of the total cost. It would be necessary to determine how the costs of the other components would change to get a complete accounting of the effects of replacing Indian Point. Some of these costs are set by regulators and are subject to change. Consequently, unlike modeling wholesale prices, there is no consistent structure for modeling the other costs, and it is virtually impossible to predict how they would change in different scenarios.
>
> The best examples of the other costs of generation are (1) payments for availability in the installed capacity (ICAP) market, and (2) payments for reserve capacity. In addition, the discussion of reliability in this chapter explains why the current structure of markets is still not providing sufficient incentives for new merchant projects. The implication is that investors will have to be paid some form of additional premium above the revenue received from the existing markets if new capacity is going to be built. In the long run, customers will have to pay for all of the additional costs of generation as well as for purchases in the wholesale market.
>
> Information on the performance of the wholesale market is readily available, but information about the other costs of generation is much more limited. Patton (2005, pp. 22-25) provides a valuable discussion of the performance of the ICAP and reserve markets; in that report, Section F, and Figure 16 in particular, shows a "net revenue analysis" of the annual net revenue (revenue minus production costs) in 2002-2004 for a combined-cycle turbine and a combustion turbine in different locations. For generators in New York City, the ICAP market is the primary source of net revenue for combustion turbines (roughly $140,000 per year per MW out of a total net revenue of $160,000 per year per MW in 2004) and a major source for combined-cycle turbines (roughly $140,000 per year per MW out of a total net revenue of $260,000 per year per MW in 2004). The net revenue from the ancillary service markets (e.g., reserve capacity) is small for both types of turbine (roughly $10,000 per year per MW). The net revenues for generators on Long Island are similar to the levels in New York City, but for upstate generators, the net revenue from the ICAP and reserve markets is very small (roughly $25,000 per year per MW).
>
> The discussion above is relevant for assessing the cost to customers of replacing Indian Point because it shows the importance of the location of capacity on the magnitudes of the "other" costs of generation. In New York City and Long Island, customers will eventually have to pay the relatively high wholesale prices for all of their purchases (the annual average prices in 2005 were $83 per megawatt-hour (MWh) and $98/MWh, respectively, compared to prices ranging from $65/MWh to $72/MWh in Zones A through F upstate) and the high other costs of generation for all generating capacity in New York City and Long Island (Zones J and K). New capacity that is built in zones other than J and K will incur relatively low costs in the ICAP and reserve markets but may require a higher premium to make them financially attractive (i.e., because the net revenue from the existing markets will be low). It is beyond the scope of this study to try to determine the net effect of these offsetting factors.
>
> The current regulatory strategy in the ICAP market is to make all generating capacity in a region eligible for capacity payments. Hence, the relatively high prices for capacity in Zones J and K are paid to all installed capacity that have offers accepted in the ICAP auctions for those zones. Nevertheless, it is probable that additional premiums will have to be paid to get new merchant capacity built.
>
> An alternative regulatory strategy is to direct capacity payments to cover the premium for new capacity, and possibly for existing capacity that operates most of the time at a minimum level but is still essential for reliability. This alternative strategy may be a less expensive way to maintain reliability in the long run, because making capacity payments to all installed capacity in the current ICAP market places no obligation on existing generators to build new capacity. Once again, there is a lot of uncertainty about how regulators will decide to deal with current concerns about reliability and what the additional costs will be above the price in the wholesale market.

this type of change in price behavior have been discussed extensively in the regulatory literature. Competitive spot prices will provide enough income to cover the operating cost of peaking capacity but not the capital cost, and as a result, the owners of peaking capacity do not earn enough in the spot market to be financially viable.

There are various ways to provide additional income to generators, but the current projections of installed generating capacity made by NYISO suggest that the market procedures adopted in the NYCA have not been entirely effective. In particular, installed capacity in the New York City metropolitan area could fall below the level needed to meet industry standards for reliability by 2008 (NYISO, 2005). Regulators had not anticipated this situation only a year ago. The outlook in 2004 indicated that sufficient new generating units had been approved and were expected to be completed in the near future so that standards for reliability in the NYCA would be exceeded for another 10 years. Subsequently, many of the proposed new generating units were delayed indefinitely, owing to the unfavorable market conditions faced by investors.

Given the size and importance of the financial, commercial, and residential sectors in the New York City region, the very high cost of blackouts makes it essential to maintain a reliable supply of electricity to customers in the region. Evidence from other published studies demonstrates that the value of avoiding a blackout is likely to be many times the typical wholesale price of electricity (Hamachi LaCommare and Eto, 2004). In other words, customers are willing to pay a substantial amount to ensure that the supply of electricity is reliable, and the current industry standard of limiting outages to less than 1 day in 10 years, established by the North American Electric Reliability Council (NERC), is consistent with this high value of reliability (NERC, 2004). The possibility that reliability in the New York City region will fall below the industry standard by 2008 presents a challenge that regulators will have to address in the near future (NYISO, 2005).

Before new ways are considered to supplement the earnings of generators in the spot market, it is important to identify three assumptions that have been adopted by regulators in the NYCA, which have limited the effectiveness of market forces in maintaining reliability, as explained in Appendix E. These assumptions, which are consistent with the NYISO planning strategy,[2] are

1. That setting minimum levels of installed generating capacity is an acceptable proxy for meeting the NERC standards for reliability in the NYCA,
2. That setting locational requirements for generating capacity in New York City and Long Island is an acceptable way to offset the limitations of the legacy transmission system into the New York City region,[3] and
3. That the political realities in the NYCA make it infeasible to allow high price spikes in the spot market above short-run competitive levels as a way to supplement the earnings of generators.

By accepting the first two assumptions, regulators have reduced the problem of determining how to maintain the reliability of supply to one of simply ensuring that requirements for installed generating capacity in New York City and Long Island, and the reserve margin requirement for NYCA, are met. Clearly, this transformation of concerns about the reliability of supply to concerns about minimum levels of generating capacity (generation adequacy) is more likely to be economically efficient when the transmission system is relatively robust and the availability of generating capacity is the main limiting factor. This is no longer the case in the NYCA, given the structure of the legacy transmission system and the size and location of New York City. Nevertheless, regulators have accepted the assumption that meeting locational requirements for generating capacity is an effective strategy for meeting the NERC reliability standards. By focusing on generation adequacy, however, the current regulatory practices followed in the NYCA, using the NYISO planning models adopted in Chapter 5, estimate the required levels of generating capacity. This modeling framework tends to discount the potential value of upgrades to the transmission system as a way to improve the reliability of supply. However, alternative planning models could be adopted that, in principle, would treat generation and transmission in a more integrated way. The development of such models was beyond the scope of this analysis.

By adopting the third assumption—that it is desirable to maintain short-run competitive spot prices—regulators have ensured that earnings for some peaking units that are needed for operating reliability will be insufficient to make them financially viable.

Two distinct ways to address the economic problem of funding sufficient capacity are under discussion. The first is to supplement the profits earned in the spot market for all generating units by providing enough additional income from another source to cover the "missing" capital costs. The second is to use targeted contracts, such as Power Purchase Agreements (PPAs), with sufficient generating units to meet reliability standards.

Regulators in the NYCA have chosen the first approach, because they apparently consider that it is economically fair for both the owners of installed generating capacity and po-

---

[2] The assumptions follow from NYISO comprehensive reliability planning and the NERC reliability criteria (NYISO, 2005).

[3] System security planning using the so-called N-1 analysis for generation and transmission failure could be applied as an alternative planning approach.

tential investors in new capacity. In contrast, contracts with some but not all generators are inherently discriminatory and may distort market behavior. Although the basic rationale for these arguments is consistent with regulatory theory, there is still no guarantee that the approach chosen by regulators for maintaining reliability in the NYCA will be either effective or economically efficient.

In other electricity markets (e.g., Australia), short-term price spikes in the spot market are acceptable to regulators so long as the average spot prices are competitive. Discussions are under way in Texas on adopting a similar approach. The regulatory focus in this type of market is on maintaining long-run competitive prices, rather than short-run competitive prices, and the effect is to make the earnings of generators correspond more closely to the true costs of production, including the capital costs. In the NYCA, however, regulators appear to try to avoid high price spikes in the spot market. Given this restriction, one possible way to recover the missing capital costs for peaking units is through a separate market for generating capacity.

The approach just described has been proposed by regulators in the three northeastern power pools. At this time, NYISO is the only one of the three to fully implement such a capacity market. There is still a considerable amount of political opposition to the proposal in New England, and there is an ongoing debate about it among stakeholders in the "Pennsylvania Jersey Maryland" (PJM) power pool. To provide a perspective on current conditions in the NYCA, it is important to understand why there is so much controversy about the effectiveness of capacity markets as a way to provide the incentives needed to initiate merchant investments in new generating capacity.

Initially, the installed capacity (ICAP) market run by NYISO was simply an auction for availability, designed to ensure that enough installed generating capacity would be available to meet the projected loads in New York City, Long Island, and the NYCA for a few months ahead. In general, this type of ICAP market does provide additional earnings for generators; these earnings may be significant for the continued financial viability of some peaking units. On the one hand, for example, the existence of the ICAP market may result in some units being available instead of unavailable, and it may also delay the retirement of some units. On the other hand, the extra earnings from the ICAP market are really a bonus for other generating units, such as nuclear and hydro units, because these units would be available anyway without the ICAP market. Nevertheless, regulatory theory implies that all generators should be eligible for participation in the ICAP market, and this issue is not the major source of controversy among regulators.

The main controversy about the ICAP market arises when the objectives of this market are extended to deal with the construction of new generating capacity. The following three limitations of an ICAP market in providing incentives for potential investors are explained more fully in Appendix E:

- The time horizon in an ICAP market does not extend far enough into the future to meet the needs of investors.
- It is unrealistic to place the primary responsibility for maintaining generation adequacy (and by assumption, system reliability) on load serving entities (LSEs).
- There is no legal requirement that any of the additional earnings from an ICAP market be used to build new generating capacity when and where it is needed.

The basic structure of the ICAP market in the NYCA is that regulators have placed a legal obligation on buyers (LSEs) to purchase enough generating capacity to meet their projected load plus a reserve margin before the spot market for electricity clears. (LSEs can also meet some of their own capacity requirements if these sources are certified by NYISO.) The final monthly auction in the ICAP market clears a few days before the month begins. It represents the last chance for LSEs to meet their capacity obligations without paying a substantial penalty.

The final monthly ICAP auction includes a specified "demand curve" that is designed to ensure that the market price of capacity is equivalent to the capital cost of a peaking unit if the total supply of capacity in the ICAP auction falls to the minimum amount needed to meet the regulated standards of generation adequacy. The market price will be higher (lower) if the total capacity offered is lower (higher) than the required amount. The basic objective of the current ICAP market is to make the market price of capacity cover the missing capital cost of a peaking unit when the market is economically efficient (i.e., when the total supply of capacity is equal to the amount needed for adequacy).

The financing of new generation and transmission facilities in the NYCA, whether it is needed to accommodate the retirement of existing facilities, the projected growth of load, or the intentional shutdown of Indian Point Units 2 and 3, must be understood in the context of the current hybrid mix of competitive markets and regulatory interventions that has resulted from the restructuring of the electric sector. Proposals to build new generation and transmission facilities are no longer preapproved by the New York Public Service Commission with the implicit guarantee to investors that all prudent production costs and capital costs will be recovered from customers. Investors face "regulatory risk" due to concerns that current market rules may be changed in the future, as they were after the energy crisis of 2000 and 2001 in California, as well as competitive risk. Risk increases the financial risk of an investment in new generating capacity, implying that the cost of borrowing capital for investors will be substantially higher than it would be under regulation.

Market forces have been able to maintain adequate levels of generation with relatively little regulatory intervention in Australia, for example, but not in the NYCA. Appendix E explains why the successful efforts of regulators to maintain short-run standards of economic efficiency in the spot market have undermined the financial viability of generating

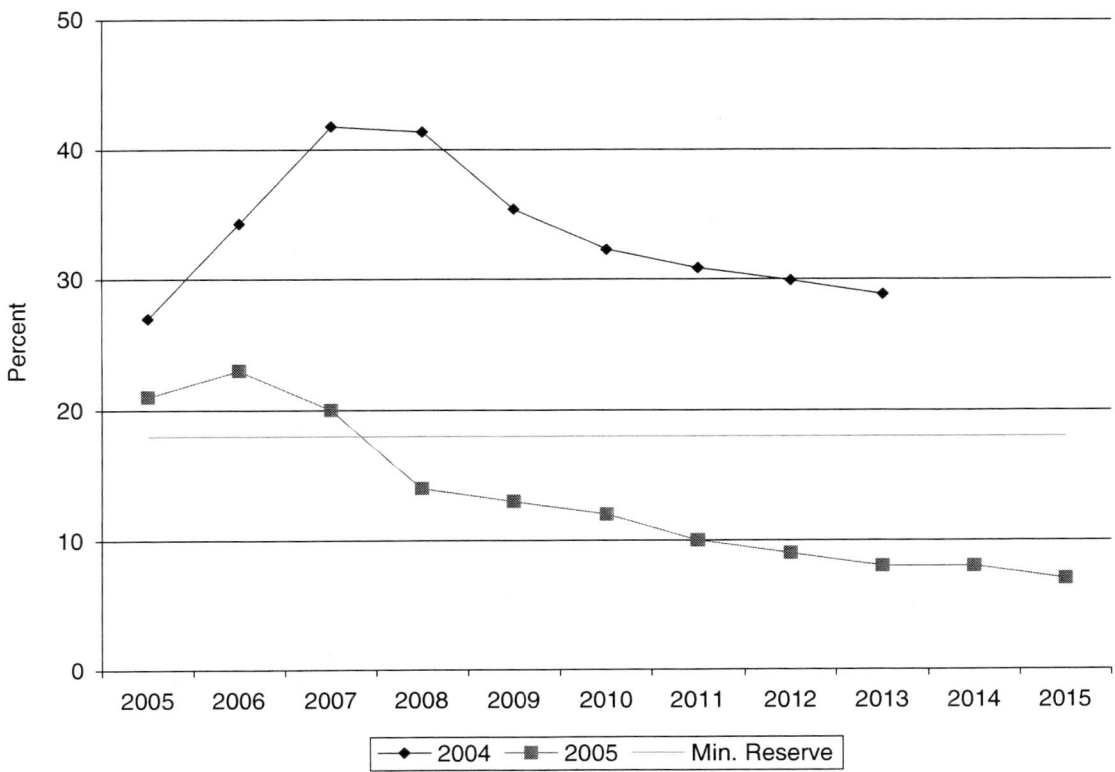

FIGURE 4-1 Projections made by NYISO in 2004 and 2005: summer reserve margin for generating capacity in the New York Control Area.
SOURCES: Projections made in 2004 from NYISO (2004), Table V-2; those made in 2005 from NYISO (2005), Table 7.2.1.

units that are needed for reliability (i.e., units with low capacity factors). This change in the pattern of spot prices has reduced the earnings of peaking units relative to baseload units and, coupled with the current uncertainty about the future prices of fossil fuels such as natural gas, has led to delays in the construction of new facilities already licensed in the NYCA.

The deteriorating outlook for reliability in the NYCA is best summarized by the drop in projected reserve margins for generating capacity from the forecast made in 2004 to that in 2005. A year ago as of this writing, in 2004, the reserve margin in 2008 was expected to be over 40 percent; however, the 2005 projection for 2008 was less than the 18 percent needed to meet the NERC reliability standards.

Figure 4-1 shows the two projections of reserve margins for the summer peak load in the NYCA that were published by NYISO in 2004 and 2005. The drop in the projected reserve margins shown in the figure was caused by delays in the construction of new generating units that had already received construction licenses. The lists of potential new generating units underlying the two projections of reserve margins in 2004 and 2005 are essentially the same, but the "Proposed In-Service" dates are quite different. In 2004, 2,038 MW were under construction (four units); 3,120 MW were approved (seven units); and 1,605 MW had applications pending (two units), for a total of 6,763 MW. Five of the nine projects (2,430 MW) with applications approved or pending had proposed in-service dates no later than 2007. However, although the amount of capacity under construction was still 2,038 MW in 2005, none of the other nine projects had proposed in-service dates, and under current market conditions, there is no guarantee that any of these generating units will actually be built.[4]

The current concern about meeting the levels of generation adequacy needed to maintain reliability in the NYCA coincides with two important changes in regulatory procedures and responsibilities. First, a new Comprehensive Reliability Planning Process (CRPP) was implemented by NYISO in 2005; the new forecasted reserve margins for 2005

---

[4]The time frame for deciding on alternatives is not known. However, NYISO is sufficiently concerned about the delays or cancellation of new generation capacity to have requested proposals for alternative solutions for addressing electricity supply, especially for the New York City area as discussed in Chapter 5.

shown in Figure 4-1 were produced for the CRPP. The second regulatory change is that the Energy Policy Act of 2005 has given FERC stronger oversight responsibilities for maintaining reliability standards for all users of the bulk power system in the United States. Under this legislation, FERC is permitted to pass these responsibilities to a single Electric Reliability Organization (ERO) that will determine explicit reliability standards and also have the authority to enforce them.

When uncertainty about the retirement dates of existing generating units in the NYCA is combined with uncertainty about whether new generating units will be built, the task of ensuring that there will be enough installed generating capacity to meet reliability standards is very challenging. Nevertheless, reliability standards must be met because the cost of blackouts in a dense urban area like New York City is so high. Although the importance of maintaining reliability has been recognized in the implementation of the CRPP and the Energy Policy Act of 2005, it is still too early to know exactly how regulators will meet their new responsibilities and use their new authority. Nevertheless, it is clear that the objective of meeting reliability standards is a high priority at both the state and federal levels, as it should be.

The current pessimistic outlook for maintaining reliability standards in the NYCA also poses a challenge for this committee. Although the committee is convinced that regulators should place the highest priority on maintaining reliability, the committee's responsibilities do not include making specific recommendations about how this should be done. Since the current projections of installed generating capacity fall short of the minimum levels needed for generation adequacy, the first step in evaluating alternatives to Indian Point is to specify a new scenario that does meet reliability standards with Indian Point operating. The assumptions used to specify this scenario are discussed in detail in Chapter 5 of this report.

The Permitting Process with Article X

The committee is aware that New York State will face a formidable task in constructing sufficient power plants to satisfy the continued load growth being experienced in the state and to replace old power plants that are to be retired for various reasons. Early retirement of Indian Point would add to those problems, whichever options are selected. A business-as-usual approach is unlikely to achieve the additional capacity that would be required. The siting of new major electric generating facilities would be facilitated if the State of New York reauthorized Public Service Law Article X, which expired on January 1, 2003.[5]

Article X had centralized the process of environmental permitting for electric power plants and provided for a firm, finite schedule for the approval or denial of environmental permits, limiting the risks of delay. This approach grew in importance with the restructuring of the electric power sector. Before restructuring, the monopoly franchise utility would propose a project based on the need to meet local loads, and the appropriate regulatory body (e.g., the NYPSC) approved or denied the proposal. In this approach, additional costs imposed on the utility company by environmental regulatory requirements or delays could be (and usually were) passed on to ratepayers. Now, the costs and risks of power plant development fall to private developers, who seek to be compensated in the marketplace—which may be intolerant of any additional expenses due to delays or other contingencies.

While it was in force, Article X set forth a review process for consideration of applications to construct and operate electric generating facilities of 80 MW or more. An approval would result in the applicant being granted a Certificate of Environmental Compatibility and Public Need, which is required before the construction of such a facility.

Most of the review under Article X is conducted by two examiners, one from the New York Department of Public Service and one from the New York Department of Environmental Conservation (NYDEC). Numerous opportunities for public involvement in hearings and other proceedings existed, and the applicants were required to pay fees that interveners could use, with permission of the examiners. Municipalities and individuals within a 5-mile radius of the proposed facilities were granted routine intervener status.

Within a year of receipt of the application, the Board on Electric Generating Siting and the Environment was required to make a decision. This board consisted of the chair of the New York Public Service Commission, the chair of the New York State Energy Research and Development Authority (NYSERDA), the commissioners of NYDEC, the New York Department of Health, and the New York Department of Economic Development, plus two public members who reside near the proposed facility and are appointed by the governor.

For example, in 2000 the board granted the Athens Generating Station a certificate (Board on Electric Generating Siting and the Environment, 2000). Topics that the board considered included the legality of the application and review process, regional and local aquatic impacts (including erosion control and deposition of pollutants), the visibility of the plant and stacks to the public (especially from historic sites), the visibility of the proposed cooling-tower plume, air quality, terrestrial biology, chemical storage and waste management, impacts on agricultural lands, noise, traffic, land use (including wetlands mitigation), public interest concerns (including the enhancement of competition, alternative sites, electrical interconnection, and local taxes), and the status of

---

[5]For additional information, see http://www.dps.state.ny.us/articlex_process.html. Accessed January 2006.

required permits. During the process, many interveners participated; they and the applicant agreed to many changes in plant design, some of which were fairly expensive. Important changes included shorter stacks, the use of dry cooling, the use of state-of-the-art emissions controls, and payments to mitigate various impacts. The board also imposed several conditions on the applicant in its approval.

Since the expiration of Article X, electric generating project developers must obtain all of the appropriate local and state permits and approvals and must undergo environmental review subject to the State Environmental Quality Review Act (Article 8 of the Environmental Conservation Law). Project developers may also obtain a Certificate of Public Convenience and Necessity, based on the traditional approach to adding electric generating capacity. New York's Governor George Pataki and several state legislators have proposed new laws to replace Article X, but there is none currently in place.

Industry groups (e.g., the Business Council of New York State) have promoted a new siting law, while some advocacy groups (e.g., the New York Public Interest Research Group) have expressed concerns. One specific concern is about whether or not the local community must give its permission for a new plant. Under Article X, municipalities could participate in the process, but the final decision was made by the board.

If action is taken to reauthorize Article X, the following issues, among others, could be considered:

- The addition of modifications and measures to Article X's procedural requirements that would enable the siting board to streamline its review when interested parties, including affected communities groups, had reached a consensus as to the specific issues presented by an Article X application.
- The appropriateness of developing specific procedures with respect to the expansion, modification, or repowering of existing major generating facilities.

In addition, the committee suggests consideration of the reauthorization of Article 6 of New York's energy law, for statewide energy planning, that expired on January 1, 2003.[6] In addition to statutory modifications, the following administrative steps might be taken:

- The Energy Planning Board could meet annually to coordinate the development and implementation of energy-related strategies and policies, receive reports from the agencies' staffs on the compliance of major energy suppliers with its information-filing requirements, and receive summary reports on the information filed.
- The information-filing regulations of the Energy Planning Board could be modified to recognize new entrants into the energy marketplace and the need for pertinent energy-related information and data.

SOCIAL CONCERNS

The social concerns considered here are environmental impacts, energy security, and indirect socioeconomic factors, including impacts on the affected communities. The concerns can have a significant effect on what sort of facilities can replace Indian Point and where they can be built.

Environmental Regulation

All energy technologies have environmental impacts. Replacement technologies discussed in Chapters 2 and 3 include efficiency and distributed generation,[7] natural-gas-fired turbines, and, potentially, coal-fired generation (any new coal plants are likely to be upstate or out of state, with long-distance transmission). Replacing the Indian Point nuclear power generators with a different type of electricity supply may reduce some environmental effects but may increase others. In contrast, energy-efficient technologies reduce the need for both capacity (megawatts) and energy (megawatt-hours) and thus tend to reduce environmental impacts (unless their manufacture, recycling, or disposal is problematic).

In New York as elsewhere in the United States, a complex set of regulations and permit requirements are in place to manage these effects and to ensure that they impose a minimal burden on the public and the environment. Environmental effects of nuclear power plants associated with plant construction, fuel production, and disposal of radioactive waste have been evaluated extensively elsewhere (e.g., McFarlane, 2001; NRC, 2001, on spent fuel disposal) and are outside the scope of this study. In normal operation, nuclear power plants such as those at Indian Point emit very little air pollution. Large releases of radionuclides might occur as the result of an accident or attack (Farrell, 2004b), but that potential has a relatively low probability. Indian Point does have a significant impact on the Hudson River, as discussed in the subsection below, on "Water Use."

The most significant pollutants from natural-gas combined-cycle plants, the most likely fossil-fueled generation replacement for Indian Point, are nitrogen oxides, NO and NO_2 (designated as NO_x), and, to a much lesser extent, car-

---

[6]Article 6 concerns the organization and functions of the state Energy Planning Board.

[7]On-grid renewable generation options were also considered, but the committee determined that they were not competitive in the timescale of the study.

bon monoxide (CO), volatile organic compounds (VOCs), and particulate matter (PM) (e.g., Barboza et al., 2000). However, emissions of all of these pollutants are sufficiently low from gas turbines or can be controlled sufficiently well so that it is quite feasible to obtain air quality permits which guarantee plant operation that protects human health and the environment (U.S. EPA, 1997). Carbon dioxide emissions, currently not regulated, are discussed below.

The effect of possible replacements for the Indian Point reactors on a broader size range of particulate matter (PM_{10}) emissions is likely to be small because of (1) permitting requirements that will require low emission rates and a tall stack to control local effects, and (2) emission-reduction offset requirements that will yield a net decrease in regional emissions of PM_{10}. For the more important emissions of the smaller particulate matter (called $PM_{2.5}$), the effect on mass emissions is largely determined by SO_2 and NO_x emissions, which, on a regional basis, will be unaffected owing to the emissions caps imposed on the electric power sector for these pollutants.

Three important pollutants from power plants, including coal-fired units, are or will be controlled by cap-and-trade programs: NO_x, sulfur dioxide (SO_2), and mercury (Hg) (U.S. Congress, 1990; Farrell, 2004a).

Both NO_x and SO_2 can have *direct* negative effects on human health, and so are "criteria pollutants," with their own standards under the federal Clean Air Act. Southeastern New York (and, in fact, the entire country) has attained healthful air quality for NO_x and SO_2 and is classified as "in attainment" of the National Ambient Air Quality Standards (NAAQS) for these pollutants. Nitrogen oxides and SO_2 contribute *indirectly* to two other criteria pollutants, ozone (O_3) and particulate matter. The former is produced in the atmosphere through photochemical reactions of NO_x and VOCs. The latter involves nitrate and sulfate formation from oxidation of the two gases in the air forming condensable material as PM. Measured O_3 and $PM_{2.5}$ concentrations in various cities have resulted in local nonattainment of the NAAQS for these pollutants, including cities in some parts of southeastern New York. The nonattainment designation requires the state to provide plans for achieving attainment, which in turn requires reductions in NO_x and SO_2 concentrations well below levels otherwise required. These requirements affect choices of power plant technology using fossil fuels.

The attainment of the NAAQS for NO_x (as NO_2) and SO_2 has been achieved locally through the use of cleaner fuels, improved combustion technologies, and combustion by-products emitted well above ground level, to disperse and dilute remaining emissions. As with PM and CO, the regulatory process to approve new power plants involves atmospheric modeling to set emissions limitations and stack heights in order to help ensure that there are no local health impacts from the expected NO_x and SO_2 emissions. A new power plant would also be required to offset its emissions and retire emission "credits" equal to 30 percent of those emissions, creating a net reduction in regional NO_x and SO_2 emissions.

Nitrogen oxides and SO_2 contribute not only to local issues, but also to larger-scale (regional) environmental problems of tropospheric ozone, fine particulate matter ($PM_{2.5}$), acidification of sensitive ecosystems, and (in the case of NO_x) eutrophication (Regens, 1993; Chameides et al., 1994; Jaworski et al., 1997; Tucker, 1998; Solomon et al., 1999; U.S. EPA, 2000; Mauzerall and Wang, 2001; Streets et al., 2001; Farrell and Keating, 2002; Creilson et al., 2003). In order to manage these regional problems, additional controls for NO_x and SO_2 are superimposed on controls designed to ensure local air quality. These regional air-quality-related problems result from aerometric phenomena that occur over several hundred kilometers and can take several days to complete. Therefore, projecting the impact of potential fossil-fueled replacements for Indian Point requires placing them into a context of regional changes in emissions, not simply the localized changes near new power plants or urban settings.

In the United States, SO_2 and NO_x emissions from large electric generators are regulated by a "cap-and-trade" system; this type of regulation has been proposed for Hg as well (Farrell, 2004a). Current regulations for SO_2 and NO_x are contained in the Clean Air Interstate Rule (CAIR), which was published in its final form in March 2005 and will be implemented fully by 2020 (U.S. EPA, 2005).[8]

The CAIR will lower SO_2 emissions from the electric power sector across a 28-state region (including New York) by about 65 percent and NO_x emissions by about 50 percent. However, the CAIR imposes an annual cap on NO_x emissions, while the key problem in the northeastern states is summertime ozone and fine particulate formation. Some analyses suggest that the annual cap in the CAIR may not be sufficient to maintain current summer air quality in the New York area, and that an additional, seasonal NO_x control program may be required (Palmer et al., 2005).

The Clean Air Mercury Rule (CAMR) is still under review. Even without it, Hg emissions are expected to decline as a co-benefit of the more stringent controls on SO_2 and NO_x emissions.

In considering a potential replacement of the Indian Point reactors with fossil-fuel generation, the key feature of cap-and-trade systems is that emissions are limited in absolute magnitude and do not respond to changes in the amount of electricity generated or in the technologies used. While increased generation at an existing power plant may lead to additional emissions at that facility, such increased generation would not be allowed if new emission controls are added to the plant, as is happening (and has been happening for over a decade) across the nation. Even if no new control technologies are added, under a cap-and-trade system addi-

---

[8] See www.epa.gov/interstateairquality. Accessed November 2005.

tional emissions at one plant (including a new one) must be compensated for by reduced emissions from another plant. This trade-off would result in no net change in regional emission. The SO_2 and NO_x cap-and-trade programs are designed to solve such regional (not local) problems. These requirements are added to protect local air quality. Under the federal Clean Air Act amendments of 1990, the air quality standards that these policies are designed to achieve must protect human health with an adequate margin of safety.

Thus, if the Indian Point plants are replaced by gas- or coal-fired generators, total emissions of SO_2, NO_x, and Hg will not change (assuming that the CAMR or a more restrictive cap is put in place) and should not significantly affect human health. Instead, the spatial patterns of emissions may change slightly, and the cost of controlling emissions will increase slightly.

Local air quality in the immediate vicinity of power plants is controlled separately by environmental regulations (as discussed above). These regulations set limits on rates of emissions and require the use of tall exhaust stacks to ensure that pollutants are diluted sufficiently to avoid negative health impacts in the communities immediately surrounding the facilities under expected meteorological conditions (Davis et al., 2000; Goodfellow, 2000).

Most cap-and-trade systems, such as the one that controls SO_2 emissions, include "antibacksliding" provisions that prevent facilities from violating local air quality regulations through the use of emissions trading. Nonetheless, because the emissions of specific sources are not directly controlled by cap-and-trade programs, concerns have been raised about the possibility of "hotspots," areas of greater air pollution (or air pollution that is not lowered sufficiently) in the vicinity of some sources (Nash and Revesz, 2001). However, there is little evidence of hotspots having occurred in SO_2 and NO_x cap-and-trade programs (Farrell, 2004a; U.S. EPA, 2004). Nevertheless, local effects of emissions of toxics under a cap-and-trade program have been found to be a cause for concern (Chinn, 1999). Thus, it is reasonable to be concerned about the possibility of negative effects of Hg emissions if a coal-fired power plant replaces the Indian Point plants. However, the difficulty of finding an adequate site and of delivering coal in sufficient quantities to a location near New York City makes such an outcome unlikely in the short term (to 2015) examined in this study.

There is scientific consensus (with few dissenting opinions) that rising concentrations of greenhouse gases (GHGs) in the atmosphere have already caused perceptible changes in climate and will lead to further climate change in the future (Intergovernmental Panel on Climate Change, 2001). The impact of climate change may be significant for water resources, agriculture, ecosystems, and the incidence of catastrophic weather systems (Malmqvist and Rundle, 2002; Hayhoe et al., 2004). The most important anthropogenic GHG is carbon dioxide (CO_2), and the most important source of CO_2 is the combustion of fossil fuel.

Avoiding serious climate change impacts will require deep cuts in global CO_2 emissions. Deep cuts in return will require significant changes from current practices in energy supply and demand, because fossil fuels dominate global energy use (Hoffert et al., 1998). As a non-fossil-fuel source of energy, nuclear power may grow in importance in the future. Replacement of the Indian Point Energy Center with fossil-fueled generation could increase CO_2 emissions, the opposite of the direction necessary to avoid climate change.

There is currently no regulatory framework in the United States for controlling GHG emissions, but on December 20, 2005, Governor Pataki signed the Regional Greenhouse Gas Initiative (RGGI) Memorandum of Understanding, which committed New York State to proposing a cap-and-trade program to limit GHG emissions from the electric power sector starting in 2009. Six other states were part of this agreement: Connecticut, Delaware, Maine, New Hampshire, New Jersey, and Vermont. Fossil-fueled replacements for the Indian Point plant would emit CO_2 and would be subject to this regulation.

Costs of Emissions from New Fossil Power Plants

An upper-bound estimate of the cost of obtaining pollutant-emission allowances to cover annual emissions is calculated assuming two technologies that could be adopted as replacements for the Indian Point units up to 2018 and perhaps beyond. These are the natural gas combined cycle (NGCC) and coal-based integrated gasification combined cycle (IGCC), with the latter serving as a proxy for advanced pulverized coal with state-of-the-art emission-control technologies. The amount of energy required is assumed to be the amount produced by the two Indian Point units operating at 90 percent capacity factor for 1 year, which is about 17 million MWh. Assuming 80 percent capacity factors for the fossil-fueled plants, a total capacity of about 2,430 MW would be required.

For purposes of evaluation, nominally representative emission rate data are taken from the observed performance of Sithe Independence and Polk Stations, as given in the U.S. Environmental Protection Agency's (EPA's) database, e-grid. Two scenarios are considered: in one, CAIR and CAMR are implemented but there is no GHG emission control; the other is identical except that the RGGI baseline policy package is also implemented. Emission allowance prices for these two scenarios are taken from the September 2005 RGGI analysis (Table 4-1). The price of CO_2 allowances in the latter scenario is $1 per ton. While this is lower than the amount estimated in other policies, including that of the European Union, it nevertheless is consistent with current projections for the Northeast. Below are considered the consequences of a range of CO_2 charges, ranging from $1 per ton of CO_2 removed to $25 per ton of CO_2 removed.

The results are shown in Tables 4-2 and 4-3. The projected upper bound for the policy with GHG controls is only

TABLE 4-1 Estimated Future Emission Allowance Prices

Study	Description	NO_x ($/ton)	SO_2 ($/ton)	Hg ($/lb)	CO_2 ($/ton)
Energy Information Administration (2001, Table 4)	50%-75% reductions in SO_2, NO_x, and Hg	1,108-2,825	719-1,737	21,119-85,225	N.A.
Palmer et al. (2005, Table 14)	CAIR, CAMR, and seasonal NO_x cap	1,042	0-1,347	35,760	N.A.
Regional Greenhouse Gas Initiative (RGGI)[a]	Baseline: CAIR and CAMR	1,710	1,268	21,730	N.A.
Regional Greenhouse Gas Initiative	Reference: CAIR, CAMR, constant CO_2 emissions, 2009-2014	1,713	1,267	21,670	1

NOTE: N.A., not available. Abbreviations are defined in Appendix C.

[a]RGGI prices are based on the September 2005 analysis. See http://www.rggi.org/documents.htm. Accessed November 2005.

about $60 million per year, using the RGGI baseline price for CO_2 allowances. However, many other studies have suggested that higher prices for CO_2 allowances are likely. Holding the other allowance prices constant, adjusting CO_2 allowance prices to $10 per ton yields total annual allowance costs for NGCC of about $72 million and for IGCC of about $210 million. At $25 per ton of CO_2, these costs become about $175 million for NGCC and $450 million for IGCC.

Given the uncertainties in fuel prices, policies, and technologies, it is reasonable to expect that the cost of air emission allowances for fossil-fueled replacements for the Indian Point units would vary from a few million to ten million dollars per year if there is no GHG policy, and from ten million to possibly several hundred million dollars per year if a GHG policy is imposed.[9]

Water Use

The Indian Point Energy Center is located on the eastern shore of the Hudson River and uses three intake structures to withdraw approximately 2.5 billion gallons of water per day for cooling the reactor units in once-through heat exchangers; the water is returned to the river somewhat warmer (NYDEC, 2003, p. 8). Under the federal Clean Water Act, discharges of heat to water bodies are considered pollution and are regulated by NYDEC. In addition, the cooling-water intake systems at Indian Point contribute to significant mortality of aquatic organisms in the Hudson River estuary. For this reason the cooling-water intake system is also subject to regulation under the Clean Water Act and state regulations. These regulations require that the location, design, construction, and capacity of the cooling-water intake system must reflect the best technology available for minimizing adverse environmental impacts.

In 2003, NYDEC issued a draft State Pollutant Discharge Elimination System (SPDES) permit for Indian Point that required immediate and long-term steps to reduce the adverse impacts on the Hudson River estuary.[10] The short-term steps include mandatory outage periods, reduced intake during certain periods, continued operation of fish-impingement mitigation measures, the payment of $25 million to a Hudson River Estuary Restoration Fund, and the conduct of various studies. In the long term, NYDEC staff has determined that closed-cycle cooling is the best technology available to minimize environmental impacts of the Indian Point facility. However, the implementation of the very large, expensive modification is contingent on approval of the U.S. Nuclear Regulatory Commission (U.S. NRC) and extension of the U.S. NRC operating license for Indian Point and so is not yet certain.

Alternatives to Indian Point would likely also be required to use closed-cycle or "dry cooling" technologies that use little water. This type of cooling technology was required of the new Athens Generating Station up the Hudson River (Board on Electric Generating Siting and Environment, 2000). Small-scale generators (used for distributed generation and combined heat and power) use air cooling and thus have no significant water use.

Overall, potential replacements for Indian Point would have less impact on the Hudson River than Indian Point currently does. However, if Indian Point adds closed-cycle cooling, its impact would be reduced also.

Environmental Justice

Equity and aesthetic concerns about the impacts of electric power plants (and all energy infrastructure) are often called matters of environmental justice, which is typically

---

[9]Higher levels of costs would encourage energy-efficiency investments or replacements that emit less carbon, thus reducing the total cost.

[10]Available at http://www.dec.state.ny.us/website/dcs/eisanddp/IndianPointSPDES.pdf. Accessed November 2005.

TABLE 4-2 Annual Costs for Allowances to Replace Indian Point Generation, Without CO_2 Control (Regional Greenhouse Gas Initiative Baseline Scenario, No CO_2 Control)

	Nuclear Plant	Natural Gas Combined-Cycle Plant	Coal Integrated Gasification Combined-Cycle Plant
Capacity (MW)	2,158	2,428	2,428
Capacity factor	0.9	0.8	0.8
Generation (MWh)	17,013,672	17,013,672	17,013,672
NO_x rate (lb/MWh)	0	0.134	0.719
NO_x emissions (tons)	0	1,140	6,116
NO_x allowance cost (cost per ton: $1,710)	$0	$1,949,256	$10,459,070
SO_2 rate (lb/MWh)	0	0.025	1.55
SO_2 emissions (tons)	0	213	13,186
SO_2 allowance cost (cost per ton: $1,268)	$0	$269,667	$16,719,335
Hg rate (lb/GWh)	0	0	0.0397
Hg emissions (lb)	0	0	675
Hg allowance cost (cost per lb: $21,730)	$0	$0	$14,667,493
Total emission allowance cost	$0	$2,218,923	$41,845,898

NOTE: Allowance prices are based on September 2005 analysis of the Regional Greenhouse Gas Initiative. See http://www.rggi.org/documents.htm. Accessed November 2005. Abbreviations are defined in Appendix C.

defined as the fair treatment of all people, regardless of race or income, with respect to environmental issues. Ensuring environmental justice has been a matter of policy for the federal government for more than a decade, and in 2004 the U.S. Nuclear Regulatory Commission reaffirmed its commitment to this goal. In practice this means that "while the NRC [Nuclear Regulatory Commission] is committed to the general goals of E.O. 12898, it will strive to meet those goals through its normal and traditional NEPA [National Environmental Policy Act of 1969] review process" (President of the United States, 1994; U.S. NRC, 2004).

As a concept rooted in ideas of rights and fairness, not science and technology, environmental justice concerns are very different from the other types of issues discussed in this section. In addition, environmental justice concerns associated with energy can include a wide array of issues, because many people find electric power plants and transmission towers ugly and undesirable to live or work near. For this

TABLE 4-3 Annual Costs for Allowances to Replace Indian Point Generation with CO_2 Control (Regional Greenhouse Gas Initiative Reference Scenario)

	Nuclear Plant	Natural Gas Combined-Cycle Plant	Coal Integrated Gasification Combined-Cycle Plant
Capacity (MW)	2,158	2,428	2,428
Capacity factor	0.9	0.8	0.8
Generation (MWh)	17,013,672	17,013,672	17,013,672
NO_x rate (lb/MWh)	0	0.134	0.719
NO_x emissions (tons)	0	1,140	6,116
NO_x allowance cost (cost per ton: $1,713)	$0	$1,952,676	$10,477,419
SO_2 rate (lb/MWh)	0	0.025	1.55
SO_2 emissions (tons)	0	213	13,186
SO_2 allowance cost (cost per ton: $1,267)	$0	$269,454	$16,706,150
Hg rate (lb/GWh)	0	0	0.0397
Hg emissions (lb)	0	0	675
Hg allowance cost (cost per lb: $21,670)	$0	$0	$14,626,993
CO_2 rate (lb/MWh)	0	828	1,959
CO_2 emissions (tons)	0	7,043,660	16,664,892
CO_2 allowance cost (cost per ton: $1)	$0	$7,043,660	$16,664,892
Total emission allowance cost	$0	$9,265,790	$58,475,454

NOTE: Allowance prices are based on September 2005 analysis of the Regional Greenhouse Gas Initiative. See http://www.rggi.org/documents.htm. Accessed November 2005. Abbreviations are defined in Appendix C.

reason, there are often concerns that new power plants or power lines will lower property values. By contrast, some communities might welcome a new power plant because of the jobs and tax revenues it would bring.

Everyone uses electricity, and it must be generated somewhere and delivered in some way. Why should one community accept a power plant or transmission line when that facility will serve another community? This problem can create tensions among communities or between residents of different states. Indian Point serves Westchester County and New York City. Once the power goes onto the grid, it is indistinguishable from all other power sources, but Indian Point is basically a local plant for Westchester County and New York City. In fact, it is essentially the only generating plant in Westchester County. New York City is required to generate 80 percent of its power, but Westchester County currently has no local generation requirement. As noted elsewhere in this report, if Indian Point is closed, it will have to be replaced at least in part with new generating capacity. If these are not local plants, then all of Westchester County's power would have to be imported, impacting other communities that might object to new facilities being imposed on them.

This problem has been exacerbated by the transition from the traditional model of a regulated monopoly franchise in the electric power sector toward a model of a competitive generation market with monopoly franchise distribution utilities and a transmission system owned by various firms, but coordinated by an independent system operator. In this new framework, the traditional concepts applied to proposed power plants—including estimating the public interest in granting construction permits against the need for new generation to meet local loads—no longer fits. Instead, plants are built to be competitive in the marketplace, as embodied in the New York State Energy Plan, which describes competition as being in the public interest, as discussed earlier in this chapter.[11]

As discussed in Chapter 1, safety is a primary concern for many people living near Indian Point. They feel threatened by the plant and want it closed. This committee has not assessed the vulnerability of Indian Point. It defers to other experts to analyze whether those risks are real or negligible. What this committee can say is that the socioeconomic, environmental, and environmental justice impacts of replacing Indian Point are significant, although not universally negative. The committee also notes that safety risks of the plant would not be eliminated until the spent fuel pool is emptied, which may be many years after the plant is closed. Storage of the spent nuclear fuel, presumably onsite, may involve costs that will be borne by the current owner, or by negotiated settlement with the state or federal authorities. Policy makers must balance the risks of continued operation against the impacts inherent in closing the plant.

---

[11]See http://www.nyserda.org/Energy_Information/energy_state_plan.asp. Accessed January 2006.

Energy Security

Historically, access, availability, and affordability have dominated public policy and the design of energy systems. The costs of existing security measures have been implicitly divided between energy users, suppliers, and the government. Today, the security of energy infrastructures against deliberate attack has become a growing concern. Therefore, the context within which energy is supplied and used has evolved well past the paradigm that has led to the current physical energy infrastructure and associated institutional arrangements.

Concerns about deliberate attacks on the energy infrastructure have highlighted many critical questions to which no ready answers exist. For example: How much and what kind of security for energy infrastructure do we want and who will pay for it? Current government efforts directed at critical infrastructure protection tend to ignore this issue entirely, focusing on preventing attacks and protecting whatever energy infrastructure the private sector creates. These decisions are being made implicitly for decades, favoring certain risk-creating technologies over others (Farrell, 2004b).

Many different approaches are likely to be necessary to achieve desired levels of energy-infrastructure security. Routine security and emergency planning have obvious roles, and some features seem to inherently enhance system security, including decentralization, diversity, and redundancy. Other features, such as the utilization of specific energy sources and energy-efficiency measures, seem to have mixed effects. In particular, some renewable energy technologies can be deployed more securely than can fossil-fuel and nuclear technologies; others cannot.

Socioeconomic Factors Including Indirect Costs to the Public

The direct-cost projections, as exemplified in the scenarios discussed in Chapter 5, depend on the generation choices to replace the 2,000 MW baseload of Indian Point, the location of the generation, modifications in transmission and distribution, the timing of any projected changes, and the load growth in the New York area. Each of the options considered has certain costs associated with it in addition to the direct costs of replacement capacity and environmental protection. These likely will be borne by the public, either through arrangements with the state or through changes in the electricity rates in southern New York, although the indirect costs do not appear directly on the customer's electricity bill. At least three kinds of potential indirect, or hidden, costs are associated with replacing the power from Indian Point:

- *The economic value of the plant and its associated property.* Entergy Corporation might have to be reimbursed if the Indian Point reactors are shut down prior to their end of licenses (including the period of extended operation if they are relicensed).

- *Higher natural gas costs to all users because of increased demand from the electric power sector.* Natural gas is likely to be the main fuel for replacement generating capacity, and unless new supplies are created, constraints are likely to be experienced.
- *Changes in employment opportunities and the tax base and the loss of local services associated with the Indian Point plant.* These costs (or potential benefits, e.g., if the Indian Point plant site is converted to other economic uses) would be borne mainly by Westchester County.

The committee was unable to assess these costs, but they could be significant relative to the direct replacement costs, depending on the arrangements for the possible closure of Indian Point.

Additional sociopolitical issues to be faced by the New York communities are less tangible than are projected costs or regulation. However, there are factors that may constrain or severely limit the options for replacing Indian Point and may affect the communities in the next 20 to 30 years. These factors include the following:

- Public attitudes toward siting power plants and transmission lines (aesthetics and the not-in-my-backyard, or NIMBY, phenomenon);
- The willingness of the public to invest in energy-efficiency measures;
- Attitudes toward advanced nuclear power plants as an option that would help maintain electric energy fuel-source diversity and minimize CO_2 emissions;
- Growth and development in southern New York, requiring major decisions on resource management and infrastructure, including energy, social services, primary and secondary education, and so on; and
- Attitudes of the state government regarding the regulation of the energy sector and its approach to permitting new facilities in the state.

Accounting for these factors will influence the choices of technological options discussed or summarized in Chapters 2 and 3 in ways that are beyond the scope of this study. However, implicitly these factors, along with others discussed in this chapter, tend to reinforce the focus on the short-term options of natural-gas-supplied generation and added transmission in southern New York State as key to a replacement strategy for Indian Point.

REFERENCES

Barboza, M.J., M.J. Cannon, N.J. Charno, and P.S. Oliver. 2000. "Stationary Gas Turbines." Pp. 242-248 in *Air Pollution Engineering Manual.* W.T. Davis, ed. New York: Wiley.

Board on Electric Generating Siting and Environment. 2000. "Opinion and Order Granting Certificate of Environmental Compatibility and Public Need: Application by Athens Generating Company." Albany, N.Y., p. 127.

Chameides, W.L., P.S. Kasibhalta, J. Yienger, and H. Levy. 1994. "Growth of Continental-Scale Meso-Agro-Plexes, Regional Ozone Pollution, and World Food Production." *Science* 264(5155): 74-77.

Chinn, L.N. 1999. "Can the Market Be Fair and Efficient? An Environmental Justice Critique of Emissions Trading." *Ecology Law Quarterly* 26(1): 80-125.

Creilson, J.K., J. Fishman, and A.E. Wozniak. 2003. "Intercontinental Transport of Tropospheric Ozone: A Study of Its Seasonal Variability Across the North Atlantic Utilizing Tropospheric Ozone Residuals and Its Relationship to the North Atlantic Oscillation." *Atmospheric Chemistry and Physics* 3: 2053-2066.

Davis, W.T., A.J. Buonicore, L. Theodore, and L.H. Stander. 2000. "Introduction: Air Pollution Control Engineering and Regulatory Aspects." Pp. 1-21 in *Air Pollution Engineering Manual.* W.T. Davis, ed. New York: Wiley.

Energy Information Administration. 2001. *Analysis of Strategies for Reducing Multiple Emissions from Electric Power Plants with Advanced Technology Scenarios.* Washington, D.C.: U.S. Department of Energy.

Farrell, A.E. 2004a. "Clean Air Markets." *Encyclopedia of Energy*, Vol.1. C.J. Cleveland, ed. San Diego: Academic Press, pp. 331-342.

—. 2004b. "Environmental Impacts of Electricity." *Encyclopedia of Energy*, Vol. 2. C.J. Cleveland, ed. San Diego: Academic Press, pp.165-175.

— and T.J. Keating. 2002. "Transboundary Environmental Assessments: Lessons From OTAG." *Environmental Science and Technology* 36: 2537-2544.

Goodfellow, H.D. 2000. "Ancillary Equipment for Local Exhaust Ventilation Systems." Pp. 143-190 in *Air Pollution Engineering Manual.* W.T. Davis, ed. New York: Wiley.

Hamachi LaCommare, Kristina, and Joseph H. Eto. 2004. "Understanding the Cost of Power Interruptions to U.S. Electricity Customers." Berkeley, Calif: Lawrence Berkeley National Laboratory. September.

Hayhoe, K., D. Cayan, C.B. Field, P.C. Frumhoff, E.P. Maurer, N.L. Miller, S.C. Moser, S.H. Schneider, K.N. Cahill, E.E. Cleland, L. Dale, R. Drapek, R.M. Hanemann, L.S. Kalkstein, J. Lenihan, C.K. Lunch, R.P. Neilson, S.C. Sheridan, and J.H. Verville. 2004. "Emissions Pathways, Climate Change, and Impacts on California." *Proceedings of the National Academy of Sciences of the United States of America* 101: 12422-12427.

Hoffert, M., K. Caldeira, A.K. Jain, E.F. Haites, L.D.D. Harvey, S.D. Potter, M.E. Schlesinger, S.H. Schneider, R.G. Watts, T.L. Wigley, and D.J. Wuebbles. 1998. "Energy Implications of Future Stabilization of Atmospheric CO_2 Content." *Nature* 395: 881-884.

Intergovernmental Panel on Climate Change. 2001. *Third Assessment Report: The Scientific Basis.* New York: Cambridge University Press.

Jaworski, N.A., R.W. Howarth, and L.J. Hetling. 1997. "Atmospheric Deposition of Nitrogen Oxides onto the Landscape Contributes to Coastal Eutrophication in the Northeast United States." *Environmental Science and Technology* 31: 1995-2004.

Malmqvist, B., and S. Rundle. 2002. "Threats to the Running Water Ecosystems of the World." *Environmental Conservation* 29: 134-153.

Mauzerall, D.L. and X.P. Wang. 2001. "Protecting Agricultural Crops from the Effects of Tropospheric Ozone Exposure: Reconciling Science and Standard Setting in the United States, Europe, and Asia." *Annual Review of Energy and the Environment* 26: 237-268.

McFarlane, A. 2001. "Interim Storage of Spent Fuel in the United States." *Annual Review of Energy and the Environment* 26: 201-235.

Nash, J.R. and R.L. Revesz. 2001. "Markets and Geography: Designing Marketable Permit Schemes to Control Local and Regional Pollutants." *Ecology Law Quarterly* 28: 569-661.

NERC (National Electrical Reliability Council). 2004. "Princeton, N.J.: Long-Term Reliability Assessment." North American Electrical Reliability Council. September.

NRC (National Research Council). 2001. *Disposition of High-Level Waste and Spent Nuclear Fuel: The Continuing Societal and Technical Challenges.* Washington, D.C.: National Academy Press.

NYDEC (New York Department of Environmental Conservation). 2003. "State Pollutant Discharge Elimination System (SPDES) Draft Permit Renewal with Modification, Indian Point Electric Generating Station Fact Sheet." Albany.

NYISO (New York Independent System Operator). 2004. "2004 Load and Capacity Data."

—. 2005. *Comprehensive Reliability Planning Process Supporting Document and Appendices for the Draft Reliability Needs Assessment.* NYISO, Albany, N.Y., December 21.

Palmer, K., D. Burtraw, and J.-S. Shih. 2005. *Reducing Emissions from the Electricity Sector: The Costs and Benefits Nationwide and in the Empire State.* Albany, N.Y.: New York State Energy Research and Development Authority.

Patton, David. 2005. "2004 State of the Market Report, New York ISO." Prepared by the Independent Market Advisor to the New York ISO. Unpublished report. Potomac Economics, Ltd. July.

President of the United States. 1994. "Executive Order 12898: Federal Actions to Address Environmental Justice in Minority and Low-Income Populations." *Federal Register* 59: 7629-7633.

Regens, J.L. 1993. "Acid Deposition." Pp. 165-188 in *Keeping Pace with Science and Engineering*, M. Uman, ed. Washington, D.C.: National Academy Press.

Solomon, P.A., E.B. Cowling, G.M. Hidy, and C.S. Furiness. 1999. "Comparison of Scientific Findings from Major Ozone Field Studies in North America and Europe." *Atmospheric Environment* 34: 1885-1920.

Streets, D.G., Y.S. Chang, M. Tompkins, Y.S. Ghim, and L.D. Carter. 2001. "Efficient Regional Ozone Control Strategies for the Eastern United States." *Journal of Environmental Management* 61: 345-365.

Tucker, W.G. 1998. "Particulate Matter Sources, Emissions, and Control Options—USA." Pp. 149-164 in *Air Pollution in the 21st Century: Priority Issues and Policy*. T. Schneider, ed. New York: Elsevier.

U.S. Congress. 1990. Clean Air Act Amendments. Title 42, Chapter 85.

U.S. EPA (U.S. Environmental Protection Agency). 1997. AP-42: *Compilation of Air Pollutant Emission Factors.* Research Triangle Park, N.C.

—. 2000. *Findings of Significant Contribution and Rulemaking on Section 126 Petitions for Purposes of Reducing Interstate Ozone Transport.* Washington, D.C.

—. 2004. *The OTC NO_x Budget Program (1999-2002): Emission Trading and Impacts on Local Emission Patterns.* Office of Air and Radiation. Washington, D.C.: U.S. Environmental Protection Agency.

—. 2005. Rule to Reduce Interstate Transport of Fine Particulate Matter and Ozone (Clean Air Interstate Rule); Revisions to Acid Rain Program; Revisions to the NO_x SIP Call; Final Rule. *Federal Register* 70: 25162-25405.

U.S. NRC (U.S. Nuclear Regulatory Commission). 2004. "Policy Statement on the Treatment of Environmental Justice Matters in NRC Regulatory and Licensing Actions." *Federal Register* 69: 52040-52048.

5

Analysis of Options for Meeting Electrical Demand

The retirement of the two operating reactors at Indian Point in the 2008-2015 time frame could have significant consequences for the reliable supply of electricity in the metropolitan New York City area unless appropriate replacements are supplied. This chapter discusses the impacts that potential replacements could have on reliability, costs, and other factors.

These replacements are analyzed in the context of the current evolution of the New York electric system (the New York Control Area, or NYCA) and the regulatory system that oversees it. Until recently, the future of the NYCA was viewed with relative complacency—growth was modest, and more than enough generating plants had been proposed by developers to handle that growth. Subsequently, however, some of these plants have been canceled or deferred indefinitely. As discussed in Chapter 4, projections now show potential shortfalls as early as 2008, even without the retirement of Indian Point. Other projections, using less conservative assumptions, still predict that new capacity will be needed by 2010.

Replacing Indian Point would be likely to involve a portfolio of the options discussed in Chapters 2 and 3, including the following:

- Energy efficiency (EE);
- Demand-side management (DSM) and distributed generation (DG);
- Fuller utilization of existing generation and transmission, and deferred plant retirements;
- New generation; and
- New transmission.

The committee did not model the actions and policy initiatives that would be required to implement the supply and demand options considered here. The early-shutdown cases in particular would require some strong measures to be implemented immediately.

Different portfolios are possible, emphasizing different options. Exactly which ones would be implemented and where would make a big difference in how well the system would operate. In this chapter, example scenarios are adopted to illustrate options that could provide alternatives to the Indian Point units should they be retired.

THE NYISO STARTING POINT

The New York Independent System Operator (NYISO) recently completed the 2005 Reliability Needs Assessment (RNA; NYISO, 2005a) and the companion analysis Comprehensive Reliability Planning Process (CRPP; NYISO, 2005b). Box 5-1 briefly reviews the criteria for reliability used in the analysis. The RNA includes all generation and transmission projects currently under construction in the NYCA (2,530 MW); retirements of existing capacity currently announced (2,260 MW); and the projected electrical load through 2015. The NYISO process is described in more detail in Appendix F-1. Peak load and known NYCA resources listed by NYISO for the period under study are shown in Table 5-1.

To quantify the magnitude of the needed correction, NYISO analyzed the system adding assumed capacity where needed until adequate reliability was achieved. The Base Case in the NYISO reports is a result of analyses showing that NYCA system reliability would be determined by voltage constraints in the system due to reactive power deficiencies in the Lower Hudson Valley (LHV). In that situation, reliability falls below requirements by 2008, and an additional 500 MW would be required then, increasing to 1,750 MW by 2010.

NYISO also projects that *if* essential reactive power corrections were made in the Lower Hudson Valley, thermal transmission constraints would then control, and less generating capacity (250 MW beginning in 2009, increasing to 1,250 MW by 2010) would be required to meet NYCA reliability criteria. NYISO projected the scenario with thermal constraints controlling to 2015 (but not the Base Case), when

> **BOX 5-1**
> **Reliability Criteria**
>
> System operators generally use two main criteria for ensuring reliability: reserve margin and loss-of-load expectation (LOLE). "Reserve margin" is simply the difference between the generating capacity available to serve an area and the expected peak demand divided by the peak demand. It is measured in percent. NYISO plans for the NYCA to keep a reserve margin of at least 18 percent.
>
> LOLE is more complicated but more meaningful. It measures the predicted frequency, in days per year, that the bulk power system will not meet the expected demand for electricity in one or more zones in New York State, even if only for a short time. Equipment failures in the power system (i.e., generators and the high-voltage transmission grid together) can force part of the load on the bulk power transmission system to be involuntarily disconnected. LOLE does not include the more frequent cause of blackouts for customers that are associated with failures of the local distribution system due, for example, to falling tree limbs and ice storms.
>
> The North American Electric Reliability Council recommends a reliability standard of LOLE less than 0.1, and this standard has been adopted for the region by the Northeast Power Coordinating Council, and in turn by the New York State Reliability Council. In other words, there must be sufficient generation and transmission capability in the system that a failure to serve load somewhere in the bulk power system would be expected not more than on 1 day in 10 years. The LOLE criterion is central to the discussion of reliability in this chapter. See also Chapter 1 for a discussion of reliability.

TABLE 5-1 NYISO Base Case Peak Load and Known New York Control Area (NYCA) Resources

	2008	2010	2013	2015
Peak load (MW)	33,330	34,200	35,180	35,670
Resources (MW)	39,759	39,766	39,766	39,766
Reserve margin[a] (%)	19	16	13	12
Reserve margin[b] (%)	14	12	8	7

[a] For the calculation of reserve margin and loss-of-load expectation (LOLE), NYISO adjusted installed NYCA generating capacity downward for contracted sale of hydropower outside the NYCA and for wind power (because wind cannot be counted on during peak demand). "Resources" include the adjusted NYCA generating capacity plus Special Case Resources (SCRs, 975 MW) and Unforced Delivery Rights (UDRs, 990 MW). SCRs are agreements between NYISO and large electricity consumers (e.g., industrial companies) that will reduce load at NYISO's order. This is one of the emergency steps available to NYISO to avert outages. UDR corresponds to the two high-voltage direct current (HVDC) cables into Long Island, the Cross Sound Cable from New England (330 MW), and the Neptune Cable from New Jersey (660 MW scheduled for 2007). It is power that is expected to be available and is thus included by NYISO for planning purposes.

[b] Reserve margin without the 1,965 MW of SCR and UDR, as plotted in Figure 4-1 in Chapter 4 of this report. Assumptions on allowable resources make a large difference in the calculated reliability.

SOURCE: NYISO (2005b, p. 39).

2,250 MW would be needed. All of these projections are based on Indian Point remaining in service (NYISO, 2005a).

NYISO has solicited proposed market-based or regulated solutions from participants and stakeholders in the NYCA market. The solicitations provide that "Proposed solutions may take the form of large generating projects, small generation projects including distributed generation, demand-side programs, transmission projects, market rule changes, operating procedure changes, and other actions and projects that meet the identified reliability needs (NYISO, 2005c)."

Figure 5-1 shows projected NYCA LOLEs for the Base Case and the thermal constraint case (the top and bottom lines). It also shows two other analyses: if load increases faster than expected, and if power is constrained from flowing from upstate New York through New England and back to southeast New York. Both these assumptions adversely affect reliability to a significant extent compared to the thermal constraint case. All the analyses show that LOLE will violate the criteria limit of 0.1 in the 2008-2010 time frame.

THE COMMITTEE'S REFERENCE CASE

The committee adopted a Reference Case (with Indian Point still operating), similar to the NYISO Sensitivity Case with thermal transmission limits controlling.[1] The Reference Case includes two assumptions that differ from the NYISO case: (1) it includes constraints on the flow of power from upstate New York through New England and back to southeast New York, an assumption that NYISO did not apply in its final RNA/CRPP for the thermal sensitivity case; and (2) it used actual, though inactive, proposals for generating stations for additional capacity to meet demand, rather than NYISO's standard 250 MW plants located wherever they were needed. The committee used these as illustrative capacity additions to demonstrate the changes required to meet or exceed the LOLE requirements for balancing the electrical system. While there is no assurance that these projects will be built, presumably the developers would not have proceeded as far as they did without a reasonable expectation that the sites were viable, that fuel and transmission access would be available, and that all permits would be attainable (several have been permitted under Article X).[2] In addition, one generic plant was included, with 580 MW. Other options could be selected along with alternative timing, but the

---

[1] The committee believes that the essential corrections to reactive power would most likely be made in a timely manner, and that thermal transmission constraints would ultimately dictate system reliability and thus the additional compensatory resources required.

[2] The committee does not endorse any of these projects, nor did it analyze the financial viability of any of them; they are simply assumed to be in the generating fleet when needed in the reliability calculation. None of them is under construction. Several of them have been, or may be, canceled, although other generating companies might acquire the sites and reactivate the projects.

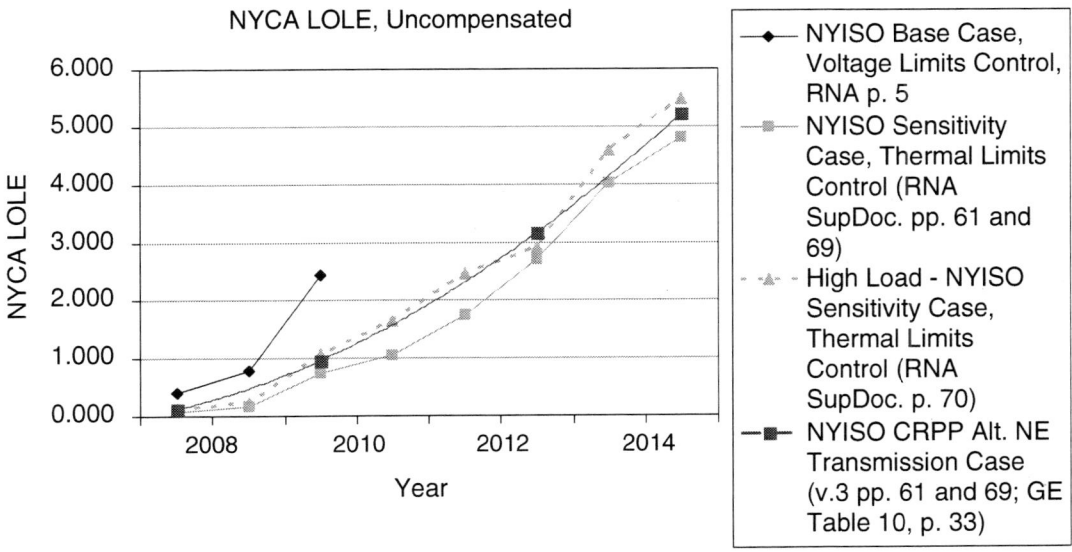

FIGURE 5-1 NYISO reliability projections. SOURCE: Derived from NYISO (2005b).

additions identified serve to illustrate the kinds of response envisaged for Indian Point replacement. The generating capacity changes assumed (beyond the 2,530 MW of generation and transmission expected to be completed before 2008) are shown in Table 5-2.

To assist the committee with the analysis, General Electric International, Inc. (GE) was retained to run its proprietary models, MARS[3] and MAPS™,[4] of the New York State and Northeast region electric systems. The MARS model (Box 5-2) is one of the principal tools used to assess NYCA system reliability. The MAPS model allows a preliminary assessment of the impact of each option studied on NYCA system operations and economics.[5] Reliability was analyzed only for 2008, 2010, 2013, and 2015, the years that the Indian Point reactors were hypothesized to be closed.

The goal of the reliability simulations was to determine the additional resources that would be required to meet reliability standards. Generating capacity was added until LOLE met the requirement of 0.1, and the NYCA reserve margin reached 18 percent.[6]

The results of the MARS analyses are shown in Figure 5-2 in comparison with NYISO's two main cases. With the committee's Reference Case assumptions, 3,300 MW are needed by 2015 to maintain reliability (LOLE < 0.1). LOLE is well below 0.1 day per year in 2008 and 2010, slightly exceeding 0.1 in both 2013 and 2015.[7] Details of this analy-

TABLE 5-2 Additional Generating Capacity Assumed in Reference Case

Project	Capacity (MW)	NYCA Zone[a]	Online Date
SCS Astoria Energy	500	J	Jan 08
Caithness	383	K	Jan 08
Long Island Wind	15[b]	K	Jan 08
Bowline Point	750	G	Jan 10
Wawayanda	540	G	Jan 13
Generic Combined Cycle	580	H	Jan 13
Reliant Astoria I	367	J	Jan 15
Reliant Astoria II	173	J	Jan 15
Total Power	3,308		

[a]See Figure 1-3 in Chapter 1 of this report for a map of the New York Control Area (NYCA) zones.
[b]FPL Energy has proposed a 150 MW wind energy project off the south shore of Long Island. Wind is an intermittent power producer, and only a small fraction of rated capacity may be available during peak load. The committee used 15 MW for this project in its reliability analysis. NYISO did not use any of the 47 MW of existing NYCA wind capacity in its reliability analyses.

SOURCE: As shown in Hinkle et al. (2005).

---

[3]GE's MARS: Multi-Area Reliability Simulation. See http://www.gepower.com/prod_serv/products/utility_software/en/downloads/10320.pdf.
[4]GE's MAPS™: Multi-Area Production Simulation. See http://www.gepower.com/prod_serv/products/utility_software/enge_mars.htm.
[5]In identifying initial reliability needs, NYISO does not conduct an economic evaluation of resources needed.
[6]The problem is considerably more complex than this. Iterative adjustments of resources assumed are needed, and the parameters to which the model is sensitive also interact with one another.
[7]In several of the committee's analyses, the rate of adding additional resources was not optimized, resulting in instances of overcompensation; projected LOLEs are thus unnecessarily low in the years prior to 2015. In further analyses, this assumption could be corrected.

> **BOX 5-2**
> **Multi-Area Reliability Simulation**
> **(MARS) Model**
>
> GE's MARS simulation software is the same system reliability screening tool approved by the New York State Reliability Council (NYSRC) and used by NYISO in its CRPP/RNA studies. MARS uses Monte Carlo simulation of the electrical generation and transmission system of the New York Control Area (NYCA) interconnected with the four contiguous electrical power systems in the northeastern United States and eastern Canada.
>
> MARS is a "transportation" model, sometimes referred to as a "bubble and stick" model, connecting generation and loads in the grid. That is, it connects the sources and sinks of power with direct-current-like flows.

sis, along with those of the scenarios below, are in Appendix F-2.

The different results (about 1 GW difference in resources needed by 2015) of the generally similar analyses by NYISO and the committee illustrate the sensitivity of the reliability analysis—and thus the additional resources needed—to differences in initial system conditions assumed. The main differences are with transmission constraints and geographic distribution of additional generating capacity.[8] The committee believes that these two cases approximately encompass the range of additional resources needed. Appendix F discusses the differences between the analyses.

REPLACEMENT SCENARIOS

With the Reference Case defined, the committee examined several cases with Indian Point closing. First, it looked at simply closing Indian Point, either in 2008/2010 (Case b1), or at the end of current license in 2013/2015 (Case c1) with no measures to compensate for the 2,000 MW capacity reduction.[9] As expected, the LOLE in both cases increased to unacceptable levels for these cases, as summarized in Figure 5-3.

The committee then analyzed cases with additional replacement resources, representing possible solutions that might arise out of NYISO's solicitation process to restore or maintain system reliability. The goal was to determine how much compensation would be necessary to maintain reliability within criteria. All of these cases included additional, aggressive programs to improve efficiency of electricity use and stronger demand-side measures to reduce peak demand. For most of them, peak demand was reduced by 300 MW in 2008, 650 MW in 2010, 800 MW in 2013, and a total of 850 MW[10] in 2015.

Additional supply was assumed to come from the proposed TransGas Energy project (1,100 MW, which was not needed in the Reference Case) in Brooklyn. Several of the Reference Case projects were accelerated as shown in Table 5-3 for Case b2 (early retirement) and Case c2 (end-of-license retirement).

The committee explored the consequences of additional scenarios, but in less detail, only looking at 2015. These included:

1. *A 1,000 MW north-south high-voltage direct-current (HVDC) transmission line running from the Marcy Substation (near Utica in Zone E) to Rock Tavern (in Zone G, south of the current transmission bottlenecks), assumed to be operational in 2012.* Cases b3 and c3 represent the early retirement and end-of-license (EOL) retirement of the Indian Point units with this HVDC cable resource in place. The inference drawn from the results is that with such a north-south transmission option, using excess power upstate and from out of state, the potential generating resource needed downstate might be reduced from 1,100 MW to 300 MW.

2. *Higher market penetration of energy efficiency and demand-side management, Cases b4 and c4, for early and EOL shutdown scenarios, respectively.* This scenario included 1,200 MW of energy efficiency and 800 MW of DSM load-reduction measures for a net 1,950 MW reduction of peak load by 2015, mainly in the New York City area. Demand would continue to grow, but at a low rate (390 MW growth compared with 2,340 MW without the EE/DSM measures). No additional capacity beyond the Reference Case would be necessary, as the additional EE and DSM measures would compensate for Indian Point. EE/DSM measures of this magnitude would require significant, aggressive early attention by the New York State government and a high fraction of all electricity users.

3. *Sensitivity to higher fuel prices.* The systems modeled were the same as in the earlier scenarios, so reliability analysis was not necessary. The committee included this analysis to estimate the approximate economic impact of higher fuel prices. The price projections used in other scenarios are lower than recent prices, and it seems plausible that gas and oil prices could remain much higher.

---

[8] Other differences in initial assumptions are estimated roughly to account for <200 MW of the 1 GW total.

[9] Note that the license for Indian Point Unit 2 expires on September 28, 2013, and that for Unit 3 on December 12, 2015. Both could still be operating through the summer peak of their last year. In particular, the absence of Unit 3 would not seriously affect reliability until the summer of 2016. However, because of the lack of a database for 2016, it was not possible to extend the analysis past 2015, so the reactors were assumed to close in January 2013 and 2015 in order to capture the impact on peak-demand reliability. In reality, an additional year would be available for replacement.

[10] Energy efficiency measures (575 MW) and demand-side management measures (300 MW) by 2015 contribute in different ways to peak reduction. The net effect of these assumptions in the model is 850 MW reduction in peak load, not the 875 MW sum.

ANALYSIS OF OPTIONS FOR MEETING ELECTRICAL DEMAND

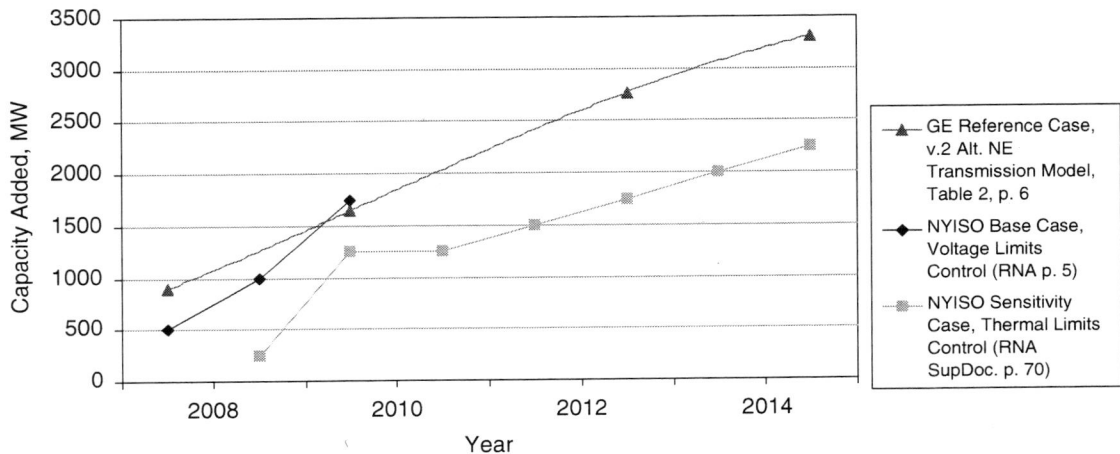

FIGURE 5-2 Approximate additional resources needed. SOURCE: Derived from NYISO (2005b) and Hinkle et al. (2005).

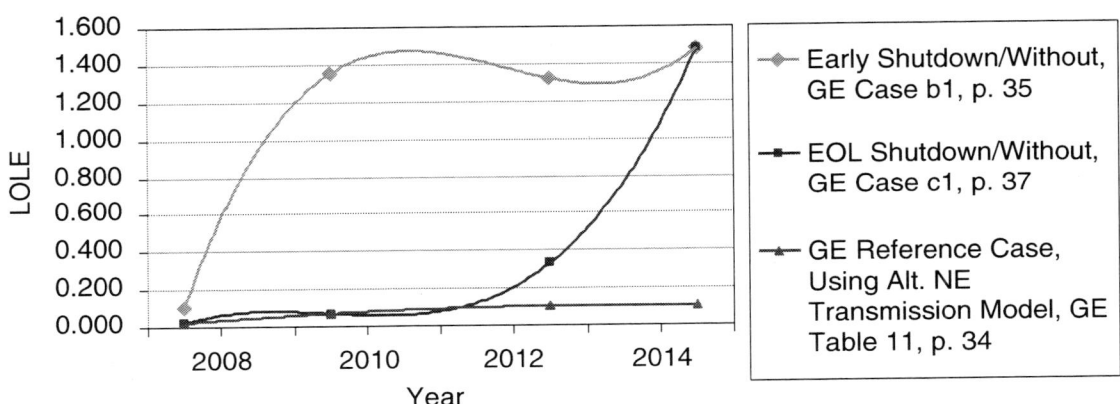

FIGURE 5-3 Impact on NYCA reliability loss of load (LOLE) of shutting down Indian Point without additional resources beyond the reference case. SOURCE: Derived from Hinkle et al. (2005).

Table 5-4 summarizes the assumed additions to resources for the various scenarios, based on achieving or exceeding the LOLE requirements. Details of the assumptions and timing of additions of illustrative resources are in Appendix F-2.

RESULTS OF RELIABILITY ANALYSES

Table 5-5 summarizes the reliability results of the cases run, showing the resulting LOLEs after compensation. Results for the Reference Case and the main cases of early and end-of-license shutdown of Indian Point are shown graphically in Figures 5-4 and 5-5, which also provide a comparison to the NYISO Base and Sensitivity Cases. Figure 5-6 shows the projected reserve margin for Case c2 (EOL shutdown of Indian Point), allowing comparison to reserve margin projections in Figure 4-1 and the impact of differing compensation.

If Indian Point is closed, roughly 2,000 MW of additional resources would be needed beyond that needed for the Reference Case. As shown in Table 5-4, the Early-Shutdown scenario (b2) requires about 4,500 MW of additional resources (total new capacity plus peak-load reduction) to be available by 2010 to meet load growth, retirements of other units, and retirement of Indian Point.[11] Of this amount, about 650 MW could result from improved efficiency and demand-

---

[11] The data on reserve margins and Figure 5-5 show the degree to which the illustrative resource additions result in overcompensation in the early years until 2013 and 2015. The schedule for adding compensation might therefore be extended in the early years.

TABLE 5-3 Capacity Additions Assumed for Cases b2 and c2

Project	Capacity (MW)	NYCA Zone	Online[a] Case b2	Online[a] Case c2
SCS Astoria Energy	500	J	2008	2008
Caithness	383	K	2008	2008
Long Island Wind	15[b]	K	2008	2008
Bowline Point	750	G	2010	2010
Wawayanda	540	G	2010	2010
Generic Combined Cycle	580	H	2013	2013
Reliant Astoria I	367	J	2008	2010
Reliant Astoria II	173	J	2008	2011
TransGas Energy	1,100	J	2010	2015
Total Power	4,408			

[a] All additions were assumed to come online in January of the year listed.
[b] See note b in Table 5-2.

SOURCE: As shown in Hinkle et al. (2005).

side management. Constructing the proposed 600 MW Cross-Hudson Cable Project, at present suspended, and extending the operation of the 880 MW Poletti 1 plant through 2010, for example, would help. Another possibility would be to extend the operation of one of the Indian Point units beyond 2010, until sufficient generation capacity could be installed in the NYCA.

In Cases b3 and c3, the added north-south HVDC transmission line was counted as a 1,000 MW resource, but the availability of sufficient generating capacity upstate was not examined in detail. As discussed in Chapter 3, the supplemental generation could come from a combination of sources, including existing or new generation upstate, or imports from Canada, all of which require additional analysis beyond the scope of this study.

This assumed HVDC line would reduce the need for new capacity in the New York City area by about 800 MW. The impact of the line on reliability would be even more substantial if (1) it would extend all the way into New York City (Zone J) and (2) if it would be backed by dedicated generat-

TABLE 5-4 Summary of Illustrative Resources Assumed to Maintain NYCA Reliability

	Year 2008	Year 2010	Year 2013	Year 2015
NYCA Peak Load, MW	33,330	34,200	35,180	35,670
NYCA Firm Capacity, MW	37,794	37,801	37,801	37,801
Total Resources with 975 MW SCR and 990 MW UDR, MW	39,759	39,766	39,766	39,766
NYISO Additional Capacity Required for Reliability, Cumulative. Thermal Limits Controlling, MW	0	1,250	1,750	2,250
COMMITTEE SCENARIOS				
Reference case, cumulative additional generating capacity assumed to meet or exceed load growth and scheduled retirements, Indian Point continues in service, MW	900	1,650	2,770	3,310
Early shutdown + compensation, Case b2, cumulative generation added above reference case, MW	540	2,180	1,640	1,100
Total Generation Added, MW	1,440	3,830	4,410	4,410
Cumulative Peak-Load Reduction by EE/DSM Measures, MW	300	650	800	850
Total Compensation for Scenario, MW	1,740	4,480	5,210	5,260
EOL shutdown + compensation, Case c2, cumulative generation added above reference case, MW	0	900	540	1,100
Total Generation Added, MW	900	2,550	3,310	4,410
Cumulative Peak-Load Reduction by EE/DSM Measures, MW	300	650	800	850
Total Compensation for Scenario, MW	1,200	3,200	4,110	5,260
ADDITIONAL SCENARIOS				
Compensation including 1,000 MW HVDC line, Cases b3 and c3, cumulative generation added above reference case, MW				300
Total Generation Added, MW				3,600
Cumulative Peak-Load Reduction by EE/DSM Measures, MW				850
Compensation including high EE/DSM measures, Cases b4 and c4, cumulative generation added above reference case, MW				0
Total Generation Added, MW				3,300
Cumulative Peak-Load Reduction by EE/DSM Measures, MW				2,000

SOURCE: Hinkle et al. (2005).

TABLE 5-5 Results of Reliability Analyses

	Year			
	2008	2010	2013	2015
NYISO 2008 CRPP/RNA Data: Table 7.3.1 Firm Resources only				
NYCA Reserve Margin, %	19	16	13	11
NYCA LOLE	0.073	0.752	2.692	4.816
For Comparison: GE-Calculated NYCA LOLE with Thermal Limits Controlling and Alternate NE Transmission Constraints	0.122	0.966	3.164	5.210
NYISO Compensation Case, with Additional Capacity as in Table 5-4. Thermal Limits Controlling				
Estimated NYCA Reserve Margin, %	19	20	18	18
Resulting NYCA LOLE	0.073	0.068	NA	NA
COMMITTEE SCENARIOS				
Reference case				
NYCA Reserve Margin, %	22	21	21	21
Resulting NYCA LOLE	0.021	0.069	0.104	0.102
Early shutdown, reference case additions only, Case b1				
NYCA Reserve Margin, %	20	16	16	16
Resulting NYCA LOLE	0.104	1.352	1.323	1.48
Early shutdown with compensation, Case b2				
NYCA Reserve Margin, %	22	24	23	22
Resulting NYCA LOLE	0.023	0.011	0.032	0.101
EOL shutdown, reference case compensation only, Case c1				
NYCA Reserve Margin, %	22	21	19	16
Resulting NYCA LOLE	0.021	0.069	0.333	1.48
EOL shutdown with compensation, Case c2				
NYCA Reserve Margin, %	18	21	18	17
Resulting NYCA LOLE	0.013	0.006	0.036	0.101
ADDITIONAL SENSITIVITY ANALYSES				
Compensation including 1,000 MW HVDC line in 2012, Cases b3 and c3				
NYCA Reserve Margin, %				19
Resulting NYCA LOLE				0.098
Compensation including high EE/DSM measures, Cases b4 and c4				
NYCA Reserve Margin, %				22
Resulting NYCA LOLE	—	—	—	0.082

NOTE: All reserve margin and LOLE results include SCR and UDR as defined in Table 5-1. SOURCE: Hinkle et al. (2005).

ing capacity. If these two conditions could be met, the transmission line would then also be counted as a resource contributing to the locational installed capacity (LICAP) requirement that Zone J's generation capacity be at least 80 percent of peak load. This HVDC line would then be analogous to the Neptune Cable now under construction, which will meet both criteria for Long Island and therefore contribute to Zone K's LICAP requirement of 98 percent.

The high levels of EE and DSM in Cases b4 and c4 would be advantageous in meeting reliability criteria, while reducing the additional generating resources required for load requirements with the retirement of the Indian Point units. Reducing demand growth by 1 MW would mean avoiding the need to build 1.18 MW to meet the NYCA reserve margin requirement. Even so, replacing the 2,000 MW from Indian Point would require reducing peak load by 1,700 MW by 2015, a very ambitious goal. The technical potential is there, and current programs are having considerable success, but progress comes in small increments that must be implemented by many people. It should be noted that the results of such programs are harder to verify than the contribution of a new generating capacity.

Corrections to reactive power are also required. The capital cost of static VAR compensation (SVC) is in the range of $50 per kilovar (kVAR), and that of a synchronous condenser about $35/kVAR (O'Neill, 2004).[12] Equipment to replace the reactive power that Indian Point is capable of supplying would cost on the order of $30 million to $45 million. In comparison, the capital cost of a 1,000 MW power plant is on the order of $1 billion. Since the cost of correcting reac-

---

[12]O'Neill is on the staff of the Federal Energy Regulatory Commission, but was expressing his own views here.

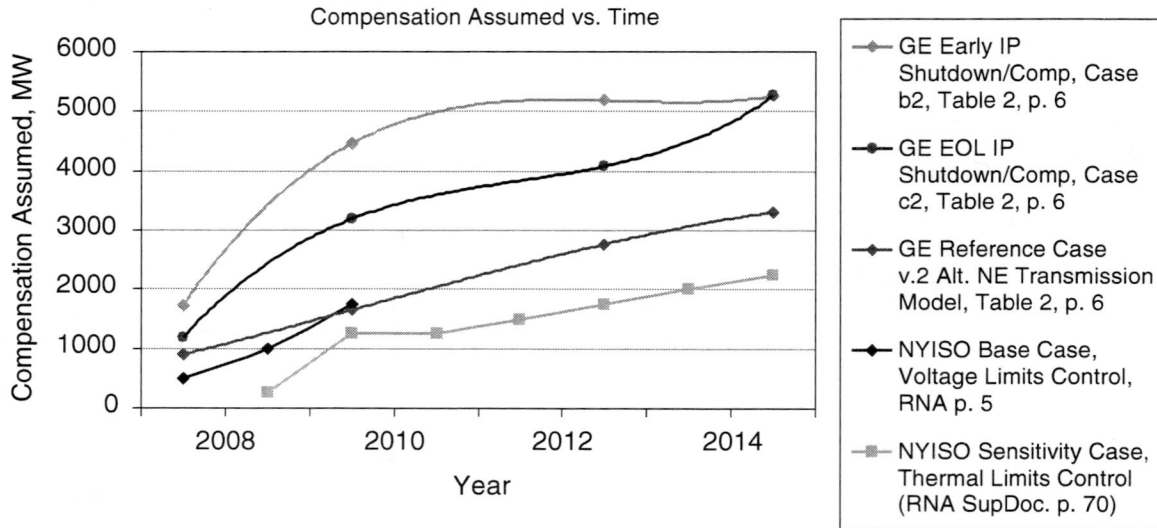

FIGURE 5-4 Capacity assumed to meet load growth and compensate for retiring Indian Point. SOURCE: Derived from NYISO (2005b) and Hinkle et al. (2005).

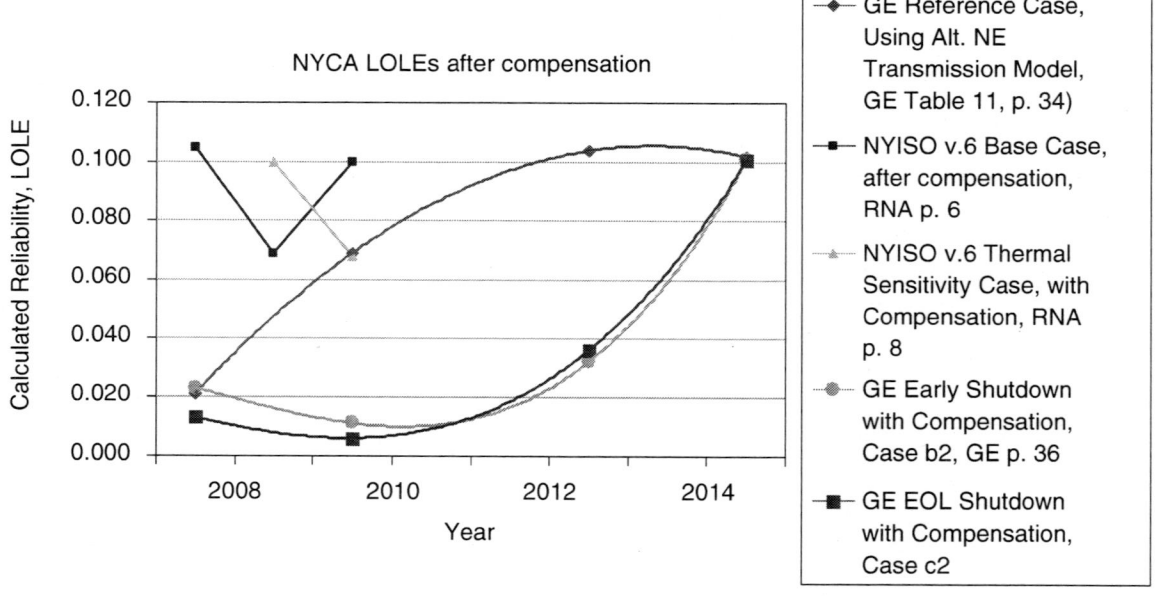

FIGURE 5-5 Loss-of-load expectation after compensation. SOURCE: Derived from NYISO (2005b) and Hinkle et al. (2005).

tive power is relatively low, the committee infers that timely local corrections to reactive power would be made.

OPERATIONAL AND ECONOMIC IMPACTS

The committee estimated the impact of closing Indian Point with the GE MAPS model for the scenarios that met reliability criteria in the MARS modeling. The NYISO case with thermal limits controlling in 2008 is the benchmark for comparing projected operational and economic impacts on (1) the diversity of the mix of fuels used to generate electricity, (2) the impact on the wholesale price of electricity, and (3) the annual variable operating cost (VOC) of producing electricity, important in the industry because it reflects the net effect of changes in both zonal generation and fuel cost (and is the fundamental variable minimized systemwide in

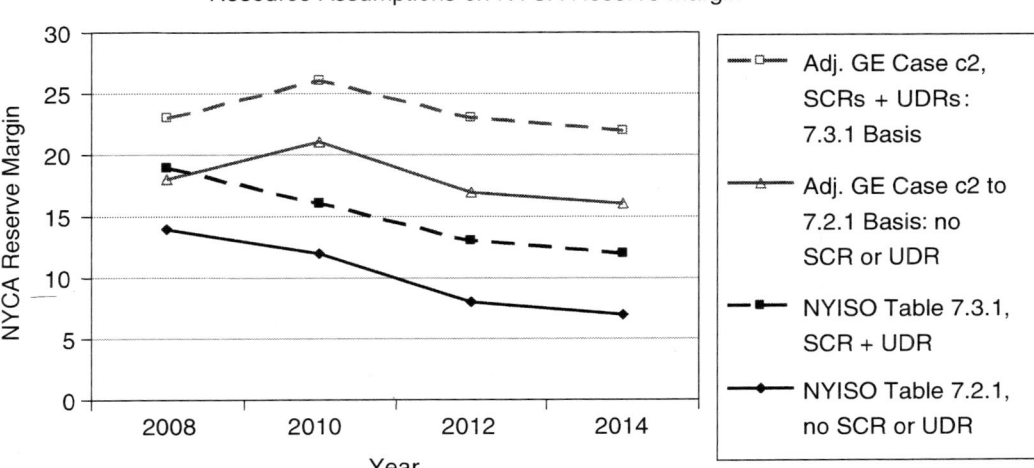

FIGURE 5-6 Projected reserve margin for End-of-License (EOL) Shutdown of Indian Point with Compensation (Case c2). SOURCE: Derived from NYISO (2005b) and Hinkle et al. (2005).

the MAPS calculations). In addition, a brief sensitivity analysis was conducted to help understand the impact that differing fuel costs would have on the cost of electricity.

Analytical Considerations

Neighboring regions (New England and part of the Pennsylvania Jersey Maryland [PJM] control area) were included in the analysis. At the outset, the committee recognized that MAPS, itself dependent on the approximate results from the MARS model analyses, would provide mainly an approximate picture of economic and cost projections into the future. Part of the MAPS model simulates the current wholesale electricity marketplace in New York State. This market is evolving to take into account aspects of pricing and investment that will differ from the present operation (see Chapter 4). Since the model cannot project such changes, confidence in the MAPS results for wholesale cost change is substantially less than in the reliability calculations of MARS.

Box 5-3 lists the main points of how the MAPS simulation works with MARS and the results produced by the simulation. Details of the modeling are contained in Appendix F-2 and the GE report (Hinkle et al., 2005).

GE's MARS and MAPS are well-accepted screening methodologies despite their many limitations. Some additional caveats are necessary in considering some limitations in the models and databases used, and thus the utility of comparisons of results for the various scenarios.

Since MAPS calculates a systemwide minimum operating cost of producing electricity, which in turn is dominated by fuel costs, the fuel prices assumed dominate the economic outputs. Fuel-cost volatility presents a significant uncertainty in interpreting the MAPS results. For the basic calculations, MAPS used a reference 2008 cost of natural gas of $5.1 per million British thermal units ($5.1/MMBtu), decreasing to $4.2/MMBtu by 2015 (both in nominal cost, or dollars-of-the-year).[13] For comparison, the U.S. Department of Energy's Energy Information Administration (DOE/EIA) reports that natural gas prices paid by electric power producers in New York State were in the range of $7.3 to $9.3/MMBtu in August 2005 (before the price increases resulting from the damage caused by Hurricane Katrina).

To assess the impact of higher fuel prices, a sensitivity study was made using a 2008 natural gas price of $7.8/MMBtu (decreasing to $7.0 by 2015). Although gas prices have dropped some in recent months, the committee recommends focusing on this case unless increased imports of liquefied natural gas (LNG) are seen as likely. Clearly, more in-depth study of gas prices and their consequences is needed.

The MAPS model of the scenarios adds considerable new NYCA generation based on modern, efficient gas-fired combined-cycle units, which require less natural gas than simple-cycle gas turbines for the same power produced. Consequently, application of these units results in lower system variable operating costs. However, no comparable assumption is made in the MAPS database for adjacent areas. This tends to lower the impact on the wholesale price of retiring Indian Point and would tend to project reduced imports of electricity from the adjacent areas in favor of increased, lower variable cost generation in the NYCA.

---

[13]Base case data set, Quarter 1, 2005, published by Platts, a Division of McGraw-Hill Companies. See http://www.platts.com/Analytic%20Solutions/BaseCase/index.xml. Accessed March 2006.

> **BOX 5-3**
> **Multi-Area Production Simulation (MAPS) Software Model**
>
> The MAPS model assesses the operational and economic characteristics of the entire interconnected region. MAPS models the electrical system in greater detail than MARS does, and is based on an economic commitment and dispatch model, also examining the flow on each transmission line for every hour of the simulation, recognizing both normal and operating reliability-related constraints. MAPS dispatches generating units in the system to meet the zonal electrical-generation requirements of a specific scenario being modeled, considering any transmission constraints. MAPS then calculates the annual variable operating cost (AVOC) of producing electricity systemwide and iterates, adjusting the dispatch of units in the system, starting with lowest variable operating cost first, to determine the minimum annual regional systemwide variable operating cost. The variable cost of producing electricity is dominated by fuel costs, but it also includes variable operational and maintenance costs, unit start-up cost (say, going from a cold start and ramping up to full electrical output), and the variable cost of emission credits consumed, where required. MAPS does not explicitly consider fixed costs, which would include capital charges; in this work, MAPS was not used to mimic the bidding strategy for bids into the wholesale market submitted by generators of electricity. Instead, pricing was equal to the variable cost of the marginal bidder, which is the theoretical limit to which economic theory drives the clearing price of a commodity in a perfectly competitive market.
>
> Having established the minimum systemwide AVOC, MAPS then provides the corresponding wholesale price of electricity, airborne emissions, and the mix of fuels used in generating electricity for each pricing zone in the system.
>
> Generation resources added to maintain reliability are inputs to the model, using MARS results as a base. MAPS does not assess the financial attractiveness of adding that capacity. It assumes that the resource is there, calculates its variable operating cost, and "dispatches" it in rank order of the variable operating cost for that resource, as capacity is aggregated to meet the then-current demand for electricity in the wholesale market.
>
> Iterative use of both the MARS reliability simulations in conjunction with the MAPS simulations for the different scenarios thus provides a basis, with some caveats, for comparing both reliability and trends of operating and economic impacts among the illustrative scenarios posed by the committee.

In evaluating the results of the MAPS analyses, readers should understand that the assumptions made tend to underestimate the projections on future wholesale prices of electricity. Therefore, the focus should be on major trends and percentage changes rather than on the absolute value of projected wholesale price of electricity. Similarly, the wholesale price of electricity modeled does not represent the final cost to consumers. Among other things, it does not include transmission and distribution costs or all of the costs for recovery of the cost of new capacity, either generation or transmission, which ultimately will, most likely, be borne by the consumer.

Fuel Diversity: Impact on NYCA Reliance on Natural Gas for Generating Electricity

Diversity of fuels used in generation is a security criterion to avoid excessive reliance on a single fuel. Generation in urban environments with minimal pollution is another criterion. New York State has benefited from ample fuel diversity in the past, and flexibility has been maintained using many gas-fired plants with dual-fuel units that can burn oil.

For the new generating capacity assumed in this study, the committee focused on natural gas in high-efficiency combined-cycle units. Natural-gas-fired generators have been the dominant choice nationwide since the mid-1980s, but that may not be strategically prudent for the next decade.

Table 5-6 compares the diversity of fuels used to generate electricity in the NYCA and the Northeast region for 2005 and 2008. Gas consumption for generating electricity is expected to increase 25 percent from 2005 to 2008. In addition, the regional shifts in fuel diversity are significant. There has been a recent reduction in the use of both oil and coal in the NYCA. In the Northeast region as a whole, the use of oil has declined, but the use of coal evidently is increasing. Finally, the projections for the Reference Case are about the same as for the Benchmark and are directionally correct in that the Reference Case adds about 1 GW of gas-based capacity and increases the change from 2005 by about another 2 percent. Further detail is shown in Appendix F-2.

Table 5-7 summarizes the projected increase of NYCA reliance on natural gas for the main options scenarios considered in this study. The table gives the percentage of NYCA reliance on natural gas for generating electricity and the impact of higher assumed fuel prices.

The MAPS projections show that reliance on natural gas would increase from 34 percent in 2008 to 44 percent in 2015 just to meet load growth and replace the capacity of units currently scheduled for retirements (the Reference Case). The projected reliance on natural gas increases to 53 percent by 2015 if Indian Point is shut down and capacity shortfall is compensated for principally by adding gas-fired units. Higher penetration of EE/DSM measures tends to reduce gas requirements, but only by about 2 percentage points. One might expect that the High EE/DSM case would lie closer to the Reference Case, but the committee was not able to investigate this further. Higher natural gas price shifts generation to other fuels, but not much, according to the MAPS projections, as the reliance on natural gas decreased only by about 3 percentage points.

In sum, the compensatory actions evaluated would significantly reduce diversity in the mix of fuels used for electrical generation in New York State. Basing compensating

TABLE 5-6 Benchmark of the Consumption of Natural Gas, Coal, and Oil for 2005 and 2008: Annual Fuel Consumption in Trillion Btu

	2005		Benchmark CRPP Thermal Case in 2008		Reference Case in 2008	
	NYCA	Northeast	NYCA	Northeast	NYCA	Northeast
Natural gas	308	804	385	1,031	392	1,032
Oil	103	132	47	59	32	44
Coal	249	2,242	218	2,344	218	2,343
Percent change from 2005						
Natural gas	—	—	25.1	28.1	27.3	28.3
Oil	—	—	−53.7	−54.8	−68.1	−66.3
Coal	—	—	−12.4	4.5	−12.5	4.5
Percent change from benchmark 2008 NYISO Base Case						
Natural gas	—	—	—	—	1.8	0.1
Oil	—	—	—	—	−31.1	−25.4
Coal	—	—	—	—	−0.1	0.0

SOURCE: Derived from Hinkle et al. (2005), plus additional personal communication with Gene Hinkle, December 2005.

resources upstate on fuel other than natural gas could lessen the reliance on natural gas, but to meet NYCA reliability criteria, that option would also require additional transmission capacity to bring power south of the congested Upstate New York-Southeast New York (UPNY/SENY) interface. Greater than 50 percent reliance on gas presents a strategic issue. In addition, it is not clear where the additional gas will be coming from. New sources, such as imported liquefied natural gas, and new transmission pipelines are likely to be required. A coal plant might be completed upstate by 2016 (the first peak-demand period after the second Indian Point reactor reaches its current EOL would be in the summer of 2016), but planning would have to start soon. Otherwise, there are few supply alternatives to gas. Considerable analysis and planning are required to develop the optimum path forward in the common interest.

Projected Impact on the Wholesale Price of Electricity

The options selected to compensate for an Indian Point shutdown would affect the operating costs for power generation. This change in turn will influence the wholesale price

TABLE 5-7 Projected Impact on Electrical Generation Based on Natural Gas for 2008 to 2015, with Sensitivity to Fuel Price

	Reference Fuel Price: NYCA Natural Gas Prices: 2008 @ $5.11/MMBtu; 2015 @ $4.24/MMBtu				Higher Fuel Price: NYCA Natural Gas Prices: 2008 @ $7.69/MMBtu; 2015 @ $7.03/MMBtu			
	2008	2010	2013	2015	2008	2010	2013	2015
Percent gas in:								
2003: 20%								
2005: 28%								
Benchmark NYISO CRPP Thermal Case in 2008	34							
Reference Case	36	38	43	44	34			
Early Shutdown with Compensation, b2	40	48	53	53	38	47	49	50
EOL Shutdown with Compensation, c2	35	39	47	53	33	37	44	50
Early Shutdown with Higher EE/DSM, b4				51				
EOL Shutdown with Higher EE/DSM, c4				51				

SOURCE: Derived from Hinkle et al. (2005).

TABLE 5-8 MAPS-Projected Impact on Electricity Wholesale Price

Case	Area	2008 ($/MWh)	2010 ($/MWh)	2013 ($/MWh)	2015 ($/MWh)
HIGHER FUEL PRICES SENSITIVITY CASES					
Benchmark of 2008 NYISO Thermal Case, Lower fuel cost		46.28			
Reference Case in Year Noted	NYCA	61	58	57	59
	Zone J	73	69	66	67
Early Shutdown with Compensation, Case b2	NYCA	63	62	60	66
	Zone J	77	75	71	79
End-of-License Shutdown with Compensation, Case c2	NYCA	60	53	58	66
	Zone J	72	60	68	79
REFERENCE CASE NATURAL GAS PRICES					
Benchmark of 2008 NYISO Thermal Limits Case	NYCA	46.28			
	Zone J	56			
Reference Case in Year Noted	NYCA	44	42	37	39
	Zone J	51	49	42	43
Early Shutdown, Case b2	NYCA	45	44	40	43
	Zone J	54	53	47	51
End-of-License Shutdown, Case c2	NYCA	43	38	38	43
	Zone J	51	43	44	51
Shutdown with HVDC Line, Cases b3 and c3	NYCA				41
	Zone J				47
Shutdown with High EE/DSM, Cases b4 and c4	NYCA				43
	Zone J				49

SOURCE: Derived from Hinkle et al. (2005).

of electricity. Table 5-8 gives the results of the MAPS-projected impact on wholesale prices of electricity in the NYCA and New York City. It is also important to recognize that other costs of producing, transmitting, and distributing electricity will ultimately be passed through, directly or indirectly, to the consumer.

As noted earlier, the committee has been unable to estimate future costs to the consumer accurately. The trends and estimated changes should be viewed as approximate. Since this is an important topic of particular importance to the consumer, additional investigation is required, including that into the evolving market structure in New York.[14] For the Reference Case results with the higher-fuel-price assumption (more likely, considering the situation today), NYCA wholesale prices are projected to remain in the range of $57 to $61/MWh between 2008 and 2015.[15] Zone J prices are consistently higher, ranging from $73/MWh to $66/MWh. If Indian Point is retired, MAPS calculates that wholesale prices by 2015 would be about $66/MWh in the NYCA and $79/MWh in New York City.

For the lower fuel prices (lower by 33 percent in 2008 and by 40 percent in 2015), the yearly average wholesale price of electricity in all of the NYCA for 2008 is projected at about $46/MWh for the Benchmark 2008 NYISO Thermal Limits case. As in the present market, there is a strong difference among zones, as the data in Appendix F-2 show in detail. The wholesale price is in the range $51/MWh to $53/MWh in Zones I, J, and K, but reaches $61/MWh in Zone H.

Some general observations include these:

- Adding substantial efficient capacity based on low-cost gas tends to lower wholesale prices in meeting load growth

---

[14] Indian Point Unit 2 was out of service for some time in 2000 as the new market was emerging and before later measures were introduced to mitigate wholesale price spikes. The NYISO Market Advisor, David Patton, analyzed the impact on wholesale prices due to the outage (Patton, 2001). During off-peak months the estimated impact on statewide wholesale prices of loss of that one unit varied from 3 to 13 percent. For summer months in the eastern part of the state, the estimated impact was as much as 30 percent. Though the market structure has changed somewhat, the impact of loss of two units could be substantial. Care should also be taken to distinguish between wholesale prices and cost to the consumer, which also includes cost of delivery to the consumer. The Westchester Public Issues Institute (2002), citing an NYPSC study, estimated that a 20 percent increase in wholesale price of electricity would translate to about a 9 percent increase in cost to the consumer.

[15] Wholesale prices are generally quoted in dollars per megawatt-hour ($/MWh). To convert to cents per kilowatt-hour (¢/kWh) divide by 10. Thus, $57/MWh is 5.7¢/kWh. Recall that these are wholesale prices. Retail prices are higher.

and scheduled retirements in both the NYCA and Zone J (which always has substantially higher prices than the NYCA). One should also recall that the unoptimized cases with compensation added more low-cost generation than needed (or is likely to be built) in the early years. Such overcompensation leads to predictions of lower wholesale prices than would result from a more realistic level of construction that just maintained reliability at a LOLE of 0.1.

- The early-shutdown scenario gives up a bit of that reduction, but not much until 2010 when Indian Point Unit 2 would be shut down.
- The HVDC case suggests the potential cost benefit of needing 800 MW less of new downstate capacity, by bringing south lower-cost electricity from upstate (assumed, arguably, to exist without new capacity upstate). It also should be noted that this case is not directly comparable to other cases, as the cost of the HVDC line would have to be passed through to the consumer in some manner, but not via the wholesale price market. The inference might still be that if no new generation is needed upstate specifically to supply the HVDC line, a lower wholesale price might well prevail downstate, but considerable analysis would be required to verify that.
- The impact of high EE/DSM penetration has only a 2 percentage point impact on wholesale price by 2015 relative to the cases with assumed EE/DSM penetration of 875 MW. This seems to be counterintuitive, and further evaluation is warranted, as this also relates to the overall incentive to invest in EE/DSM measures. In any event, it is also important to note that the ultimate cost to the consumer may be lower with EE/DSM measures, as consumers use less electricity.

An estimate of the net change in the wholesale price solely due to shutting down Indian Point, after compensating for load growth and scheduled retirements, can be obtained from GE's calculations by subtracting from the Reference Case the wholesale price estimates for the various scenarios considered. For example, by 2015 with the higher fuel prices used, the increase in wholesale price might increase $7/MWh for all of the NYCA and increase $13/MWh in New York City. For the lower-fuel-cost cases, the impact for the NYCA might be $2 to $4/MWh, and double that for New York City. However, the committee urges great caution in interpreting these numbers, since (1) the difference between two uncertain numbers is doubly uncertain; (2) it unrealistically takes shutting down Indian Point out of the context of the overall reliability situation facing New York today; (3) it allows the inference that shutting down Indian Point's 2 GW at EOL would also be compensated for by adding additional low-cost, gas-based generation; and (4) as noted earlier, the committee has low confidence in the MAPS-projected wholesale prices (based on the current locational-based marginal pricing wholesale market), which are believed to be too low.

Impact on the Annual Variable Cost of Producing Electricity

The systemwide AVOC that MAPS minimizes depends principally on the annual generation in the systemwide region under consideration and the prices of fuel there.[16] Table 5-9 gives part of the output results, providing a picture of the impacts on the AVOC for the NYCA and New York City (Zone J) in 2008 and 2015 and the sensitivity to fuel prices for the limited cases run. Values listed are the percentage changes from the Benchmark.

The data for the Reference Case in 2008 using the lower fuel prices show that AVOC initially decreases slightly, because fuel prices are low and low-cost generation is being added based on high-efficiency, natural-gas-fired units. But early shutdown of Indian Point changes this result because additional gas-based generation is added, and it has a higher variable operating cost than Indian Point, the lowest-variable-cost producer in the generating fleet—aside from hydropower. By 2015 the impact on AVOC is 21 percent higher for the NYCA and 40 percent higher for New York City. Generators of electricity there have substantially higher variable costs to cover.

The data in Table 5-9 show large impacts on AVOCs, especially in Zone J. The key points to note include these:

1. The impact of higher fuel prices is large for the entire NYCA, and especially for Zone J, with percentage increases over the Benchmark ranging from 27 to 70 percent for 2008 and from 44 to 117 percent for 2015, with the higher percentages applying to New York City. (Note that the higher-fuel-price assumptions correspond to a 50 percent increase of the 2008 price of natural gas.)

2. The AVOC in Zone J increases by 17 to 40 percent from 2008 to 2015, both relative to the Benchmark, for the Early Shutdown with Compensation scenario, because of the added capacity in Zone J.

3. Delaying the shutdown of Indian Point units until EOL shows a net early reduction in Zone J (up until 2015) because additions to capacity come later, and in the early years the impact of the use of more efficient units dominates total additions to capacity.

4. Addition of the HVDC line into Rock Tavern (Zone G) reduces the change in Zone J, as expected, as does greater penetration of EE/DSM measures. For Zone J in 2015, the combined net impact on AVOC is reduced to the range of an 8 to 14 percent increase over the Benchmark. The impact of this magnitude warrants further detailed study.

Appendix F-2 elaborates on the differing impact on AVOC in the various pricing zones, with large percentage changes

---

[16] As noted earlier, current variability in fuel prices, with bias toward higher prices than modeled, indicates that the AVOC values from the MAPS modeling are likely to be highly uncertain.

TABLE 5-9 Projected Impact on Annual Variable Operating Cost

| | Reference Fuel Prices | | | | Higher Fuel Prices | | | |
| | 2008 NYCA Gas at $5.11/MMBtu | | 2015 NYCA Gas at $4.24/MMBtu | | 2008 NYCA Gas at $7.69/MMBtu | | 2015 NYCA Gas at $7.03/MMBtu | |
Case	NYCA (%)	Zone J (%)	NYCA (%)	Zone J (%)	NYCA (%)	Zone J (%)	NYCA (%)	Zone J (%)
Reference case	−1	−2	5	−8	29	42	48	44
Early shutdown, Case b2	6	17	21	40	40	70	77	117
EOL shutdown, Case c2	−2	−3	21	40	27	40	77	117
Early shutdown, including N-S HVDC line in 2012, Case b3	—	—	12	8	—	—	—	—
EOL shutdown, including N-S HVDC line in 2012, Case c3	—	—	12	8	—	—	—	—
Early shutdown, including high EE/DSM measures by 2015, Case b4	—	—	13	14	—	—	—	—
EOL shutdown, including high EE/DSM measures by 2015, Case c4	—	—	13	14	—	—	—	—

SOURCE: Derived from Hinkle et al. (2005).

in some instances, as MAPS adjusts the electricity dispatch of various generating units to find the minimum systemwide cost. Changes of this magnitude may influence different generators of electricity substantially and could present operating and risk-management challenges, such as reliable access to fuels, and substantial shifts as new low-cost capacity is added.

Detailed results summarized in Appendix F-2 suggest an increase in AVOCs of about 10 percent for the entire Northeast region from 2008 to 2015. But this raises another caution to consider regarding the initial MAPS runs presented here and the complexity of the economic factors. The MAPS results suggest a significant, perhaps controversial, impact on regional AVOC beyond meeting load growth and compensatory actions from shutting down Indian Point. This inference might, however, only be an artifact of the calculations because of the assumptions used in the MAPS studies. Substantial gas-fired combined-cycle capacity with high efficiency is added to the NYCA over the period in question. This new capacity could be expected to displace more-expensive generation there, even older gas-fired units having lower efficiency (after compensating for the shutdown of Indian Point). However, as just one example of complexity, no comparable assumption of adding more modern gas-fired combined-cycle capacity for the New England region went into the initial MAPS model run by GE. This approach distorts the likely pattern of new generating sources that would emerge.

Sensitivity to Higher Fuel Prices

For the fuel-price sensitivity cases, the price assumptions used in MAPS differ in the following ways. For the assumed lower fuel prices, the natural gas price is 5 to 7 percent higher in PJM and New England than in NYISO; coal is 16 to 28 percent higher in New England than in either NYISO or PJM; residual oil and distillate have the same price in all three regions.[17] For the higher-fuel-price assumptions, fuel prices are the same in all regions, except that gas is 2 percent higher and coal is 16 to 23 percent higher in New England. In addition, the changes from the lower fuel prices to the higher fuel prices assume that the NYISO gas price is 50 percent higher in 2008 and 66 percent higher in 2015. The coal price is the same as in the lower set of prices; the price of residual oil rises 50 percent and 63 percent in 2008 and 2015, respectively; and the distillate fuel price goes up 38 percent and 35 percent in 2008 and 2015, respectively.

Since MAPS estimates the minimum systemwide AVOCs, these assumptions, in moving from the lower prices to the higher fuel prices, will tend to (1) slightly favor gas-based generation in NYISO over that in either New England or PJM, (2) favor coal-based generation in NYISO over coal-

---

[17]Base case data set, Quarter 1, 2005, published by Platts, a Division of McGraw-Hill Companies. See http://www.platts.com/Analytic%20Solutions/BaseCase/index.xml. Accessed March 2006.

based generation in New England, (3) favor coal-based generation slightly more in the high-fuel cases, (4) be neutral regarding gas-based generation relative to residual oil-based generation, or (5) favor distillate-based generation, relatively, except that distillate fuel is always 58 to 65 percent more costly than natural gas, so distillate-based generation penetrates only slightly in the MAPS analyses.

In evaluating the results of the MAPS analyses, it should be remembered that trends and percentage changes (rather that the absolute values of the calculated wholesale price of electricity) are mainly of interest.

COMPARING THE RESULTS WITH CRITERIA

Chapter 1 listed six criteria adopted by the committee. This section compares the results of the committee's scenario analysis with those criteria.

1. *Would the combination of demand and supply options provide adequate energy to replace that provided by Indian Point?*

A portfolio of additional supply and demand-reduction options can be identified to replace Indian Point, but they must be added to the capacity required to meet load growth and to offset generating plant retirements. The committee estimates that even if Indian Point is not retired, New York State will need about 1.2 to 1.7 GW in 2010, and 2.2 to 3.3 GW in 2015, from projects that are not already under construction. The additional 2 GW required if Indian Point were to be closed could be met by some suitable combination of new generation in the New York City area, efficiency improvements and demand-side management, and new transmission capability from upstate.

Most of the approximately 5 GW that would be needed by 2015 probably would come from new generating capacity relying at least initially on natural gas as a fuel. Energy efficiency and demand-side management have great potential, and could replace at least 800 MW of the energy produced by Indian Point and possibly much more. The new north-south transmission line analyzed by the committee also could reduce the additional generating capacity needed downstate by about 800 MW. The committee notes that critically required corrections to reactive power would have to be made locally in a timely manner, since losing the reactive power from Indian Point would only compound the projected deficiency in the Lower Hudson Valley identified by NYISO.

2. *Would the generation and transmission system be adequate to deliver the energy reliably to end users?*

Identifying the generation and transmission system capability that must be provided to replace Indian Point is much easier than determining whether it actually would get built when needed. All these measures will take time to implement, and several factors may converge to make it even more difficult. As discussed in Chapter 4, the committee questions whether the present market mechanisms are adequate to attract the capital investment required for the roughly 5 GW of new capacity and transmission corrections that would be needed by 2015. In addition, the lack of a state program, such as the former Article X, to expedite siting and licensing is likely to discourage new projects. A concerted, well-managed, and coordinated effort would be required to replace Indian Point by 2015. Replacement in the 2008-2010 time frame would be considerably more difficult, probably requiring extraordinary, emergency-like measures to achieve.

3. *How would the new combination of demand and supply options compare with Indian Point in terms of security of fuel supply for new generation?*

While the details of security comparisons are beyond the scope of this study (and would depend on the exact set of options selected), it is possible that the NYCA would be vulnerable to potential natural gas shortages. Adding several gigawatts of electrical capacity (including projects currently under construction) based mainly on natural gas supply would increase NYCA reliance on gas-based generation from 20 percent in 2003 to over 50 percent by 2015. The present gas supply and transmission capacity is inadequate to meet such future demand. In-so-far as additional gas is supplied by imported LNG, another energy security issue is introduced. Adding electrical capacity upstate based on other fuels will require additional electrical transmission capacity to serve downstate load centers, and transmission systems are inherently vulnerable to some extent. On the other hand, distributed generation has some security advantages over large generating stations. Continued vigilance at the Indian Point site for stored spent nuclear fuel will be necessary whether or not the plant is closed.

4. *How would economic costs, especially to the consumer, compare with those for continued operation of Indian Point?*

The Indian Point power plant produces baseload electricity as a low-cost wholesale provider in southern New York State. While the present "regulated competition" wholesale market depends on many factors, the projected wholesale cost without the Indian Point units, based on analysis of variable operating costs only, will tend to rise. The strongest influence on wholesale costs is fuel costs. The current volatility of natural gas prices and the structure of the wholesale market make it difficult and uncertain to project costs in 2015. In any event, it is unlikely that replacing the low-cost producer would do anything other than raise the ultimate cost of electricity to consumers.

Investors must be attracted back to the NYCA for new projects, but providing for adequate return on new capital investment will tend to increase projected wholesale prices. Costs also will increase indirectly because replacement power will increase demand for natural gas, require investment in new gas transmission infrastructure, and require expenditure for emissions permits.

5. How would environmental emissions and other impacts compare with those for continued operation of Indian Point?

Since the air emissions of New York power plants currently involve emission caps already in place, new sources would have to purchase emission rights. Thus, most pollutants would be little changed. The main change expected would be an increase in carbon dioxide (CO_2, the most important greenhouse gas) from substituting fossil fuel for nuclear fuel. If the regional plans for reducing or capping CO_2 emissions are implemented, local CO_2 increases will likely be offset with an emissions credit market. Water quality would be improved by retiring Indian Point, but much the same advantage could be achieved if the plant switched to cooling towers from the current once-through cooling.

6. What would be the impacts on local communities from closing Indian Point and replacing it with these options?

Community impacts would be mixed, depending on the choice of replacements and their locations. There would likely be potentially significant disruption in the tax base and supporting business income to Westchester and surrounding counties. A loss of employment of skilled workers would be associated with the plant's retirement. The costs of electricity are likely to rise with changes in the electrical system infrastructure in southern New York State. Projections of all of these impacts are difficult to estimate without additional information. While the committee has not studied these factors, some benefits may occur. For example, upstate communities might benefit if replacement power plants are built there. The Indian Point site could also be used for new industrial facilities that could replace the jobs and tax benefits of the nuclear station.

REFERENCES

Hinkle, G., G. Jordan, and M. Sanford. 2005. "An Assessment of Alternatives to Indian Point for Meeting Energy Needs." Unpublished report for the National Research Council. GE-Energy, Schenectady, N.Y., December 19.

NYISO (New York Independent System Operator). 2005a. *Comprehensive Reliability Planning Process (CRPP) Reliability Needs Assessment (RNA).* December 21.

—. 2005b. *Comprehensive Reliability Planning Process Supporting Document and Appendices for the Draft Reliability Needs Assessment,* NYISO, Albany, N.Y., December 21. See http://www.nyiso.org/public/webdocs/newsroom/press_releases/2005/crrp_supporting_rna_doc12202005.pdf. Accessed December 2005.

—. 2005c. Michael Calimano, solicitation letter to S.V. Lunt, R.M. Kessel, E.R. McGrath, and J. McMahon, December 22. See http://www.nyiso.org/public/webdocs/newsroom/press_releases/2005/rna_solution_letter.pdf. Accessed January 2006.

O'Neill, Richard. 2004. *Reactive Power: Is It Real? Is It in the Ether?* Harvard Electric Policy Group, Austin, Tex. December 2.

Patton, David B. 2001. *New York Market Advisor Annual Report on the New York Electric Markets for Calendar Year 2000.* April.

Westchester Public Issues Institute. 2002. *Closing Indian Point—Implications for NYC Metro Energy Supply.* June.

Appendixes

The appendixes provide information on this project and additional details and background information for the material in the report.

- *Appendix A*, "Committee Biographical Information," includes brief biographies of all the committee members.
- *Appendix B*, "Presentations and Committee Meetings," lists all the meetings that the committee held and the presenters who supplied information at the public meetings.
- *Appendix C*, "Acronyms," identifies the acronyms in the report.
- *Appendix D*, "Supply Technologies," provides additional details and background information on the generating and transmission options discussed in Chapter 3.
- *Appendix E*, "Paying for Reliability in Deregulated Markets," provides the information from which the first section of Chapter 4, "Regulation, Finance, and Reliability," was extracted.
- *Appendix F*, "Background for the System Reliability and Cost Analysis," describes the process by which the New York Independent System Operator ensures reliability and the details of the committee's analysis of future scenarios, as discussed in Chapter 5.
- *Appendix G*, "Demand-Side Measures," documents the energy-efficiency and demand-reduction technologies discussed in Chapter 2.

Appendixes D, E, F, and G were prepared by individual committee members or subgroups. They are reproduced on the CD-ROM that contains the full report but are not included in the printed report owing to space limitations.

A

Committee Biographical Information

Lawrence T. Papay (NAE), *Chair,* is currently a consultant with a variety of clients in electric power and other energy areas. Previously he held positions including senior vice president for the Integrated Solutions Sector, Science Applications International Corporation, and senior vice president and general manager of Bechtel Technology and Consulting. He also held several positions at Southern California Edison, including senior vice president, vice president, general superintendent, and director of research and development (R&D), with responsibilities for areas including bulk power generation, system planning, nuclear power, environmental operations, and development of the organization and plans for the company's R&D efforts. Dr. Papay's professional affiliations have included the Electric Power Research Institute (EPRI) Research Advisory Committee, the Atomic Industrial Forum, the U.S. Department of Energy's Energy Research Advisory Board, and the Renewable Energy Institute. He is a member of the National Academy of Engineering and the National Science Foundation's Industrial Panel on Science and Technology. His expertise and knowledge range across a wide variety of electric system technologies, from production, to transmission and distribution, utility management and systems, and end-use technologies. He received a B.S. degree in physics from Fordham University, and S.M. and Sc.D. degrees in nuclear engineering from the Massachusetts Institute of Technology (MIT).

Dan E. Arvizu is the director and chief executive of the National Renewable Energy Laboratory. He was formerly a senior vice president and chief technology officer for the Federal and Industrial Client Groups of CH2M Hill Companies, Ltd., and before that, a vice president and director of the Energy and Industrial Systems Business Group. Prior to working at CH2M Hill, Dr. Arvizu worked at Sandia National Laboratories—as director, Materials and Process Sciences Center; director, Advanced Energy Technology and Policy Center; and director, Technology Transfer Center. Dr. Arvizu was also a member of the technical staff, Customer Switching Systems, Bell Telephone Laboratories. He has experience as an executive in managing a business profit and loss, and in corporate technology commercialization as well as extensive experience in materials science applications for nuclear weapons and energy systems, and in the development of renewable energy systems, including solar thermal, photovoltaic, and concentrating solar collectors. He has been recognized for excellence in the management of technology transfer and renewable energy R&D programs. In 2004, Dr. Arvizu was appointed by President Bush to serve on the National Science Board. He received the 1996 Hispanic Engineer's National Achievement Award for Executive Excellence and has served on a number of advisory groups, including the Commercialization Advisory Board for the Solar II Central Receiver Pilot Plant. He served on the National Research Council (NRC) Committee on Programmatic Review of the Office of Power Technologies. He received his B.S. degree from New Mexico State University and his M.S. and Ph.D. degrees from Stanford University, all in mechanical engineering.

Jan Beyea is chief scientist, Consulting in the Public Interest, and is a consultant to the National Audubon Society. He consults on nuclear physics and other energy/environmental topics for numerous local, national, and international organizations. He has been chief scientist and vice president, National Audubon Society, and has held positions at the Center for Energy and Environmental Studies, Princeton University, Holy Cross College, and Columbia University. He has served as a member of numerous advisory committees and panels including the National Research Council (NRC) Board on Energy and Environmental Systems; the NRC Energy Engineering Board; the NRC Committee on Alternative Energy R&D Strategies; the NRC Committee to Review DOE's Fine Particulates Research Plan; the Secretary of Energy's Advisory Board, Task Force on Economic Modeling; and the policy committee of the Recycling Advisory Council. Dr. Beyea has been an advisor to various studies of the U.S. Congress Office of Technology Assessment. He has expertise in energy technologies and associated environmental and health concerns and has written numerous articles on the environment and energy. He received a B.A. from Amherst College and a Ph.D. in physics from Columbia University.

Peter Bradford advises and teaches restructuring and energy policy in the United States and abroad. He has been a visiting lecturer in energy policy and environmental protection at Yale University and has taught utility law at the Vermont Law School, where he is currently teaching a course on nuclear power and public policy. He is also affiliated with the Regulatory Assistance Project, which provides assistance to state and federal regulatory commissions regarding energy regulatory policy and environmental protection. Mr. Bradford was a member of the U.S. Nuclear Regulatory Commission (1977-1982). He has served on panels advising the European Bank for Reconstruction and Development on how best to replace the remaining Chernobyl nuclear plants in Ukraine and advising the Austrian Institute for Risk Reduction on regulatory issues associated with opening the Mochovce Nuclear Plant in Slovakia. He chaired the New York State Public Service Commission and the Maine Public Utilities Commission, and was also briefly Maine's Public Advocate. Mr. Bradford has written extensively on energy regulatory and energy security issues. He is a graduate of Yale University and the Yale Law School.

Marilyn A. Brown is the interim director of the Engineering Science and Technology Division at the Oak Ridge National Laboratory (ORNL). During her 22 years at ORNL, Dr. Brown has researched the impacts of policies and programs aimed at advancing the market entry of sustainable energy technologies and has led several energy technology and policy scenario studies. Prior to serving at ORNL, she was a tenured associate professor in the Department of Geography at the University of Illinois, Urbana-Champaign, where she conducted research on the diffusion of energy innovations. She has authored more than 140 publications and has been an expert witness in hearings before committees of both the U.S. Senate and the House of Representatives. She has received awards for her research from the American Council for an Energy-Efficient Economy, the Association of American Geographers, the Technology Transfer Society, and the Association of Women in Science. A recent study that she co-led (Scenarios for a Clean Energy Future) was the subject of two Senate hearings, has been cited in proposed federal legislation, and has had a significant role in international climate change debates. Dr. Brown serves on the boards of directors of several energy, engineering, and environmental organizations (including the Alliance to Save Energy and the American Council for an Energy Efficient Economy), and she serves on the editorial board of the *Journal of Technology Transfer*. She is also a member of the National Commission on Energy Policy. She has a Ph.D. in geography from Ohio State University and a master's degree in resource planning from the University of Massachusetts. She is also a certified energy manager.

Alexander E. Farrell is assistant professor in the Energy and Resources Group at the University of California, Berkeley. He is working on characterizing environmental impacts of energy production and transformation, especially air pollution and greenhouse gases, and in the economic, political, and other social aspects of energy systems with reduced environmental impacts. Previously, Dr. Farrell had been adjunct assistant professor in the Department of Engineering and Public Policy at Carnegie Mellon University and executive director of the Carnegie Mellon Electricity Industry Center. He had been a research fellow at the John F. Kennedy School of Government and at the Wharton Risk Management and Decision Processes Center, University of Pennsylvania. He also was an engineer at Air Products and Chemicals, Inc., and served as a nuclear submarine officer in the U.S. Navy. He has a B.S. degree in systems engineering from the U.S. Naval Academy and a Ph.D. in energy management and policy from the University of Pennsylvania.

Samuel M. Fleming is currently a consultant. His prior positions include executive assistant to the executive vice president for strategic planning and technology commercialization of Bechtel BWXT Idaho, LLC; senior program manager in the Operations Department of Bechtel Technology and Consulting; commercial development manager and program manager for Bechtel R&D's Cargoscan™ program; manager of the Advanced Processes Department in Bechtel R&D; project operations manager for renewable energy and fuels technologies in Bechtel R&D; manager, Process Technology Department, Bechtel R&D; manager of advanced technology planning, Fluor Engineers, Inc.; and director of technology, the Badger Company, Inc. Dr. Fleming's expertise spans a wide range in advanced technology and engineering development, economic evaluation of technologies, and project management. He has worked on various types of technology development, including advanced fuel and gas conversion, nuclear, solar, wind, geothermal, drilling, biotechnology, cargo detection, superconducting magnetic storage, and gas pipelines. He has a B.S. (Pennsylvania State University), S.M. (MIT), and Sc.D. (MIT) in chemical engineering.

George M. Hidy is principal of Envair/Aerochem. He is the retired Alabama Industries Professor of Environmental Engineering at the University of Alabama, where he was also adjunct professor of environmental health science in the School of Public Health. From 1987 to 1994, he was technical vice president of the Electric Power Research Institute, where he managed the Environmental Division and was a member of the Management Council. From 1984 to 1987, he was president of the Desert Research Institute of the University of Nevada. He has held a variety of other scientific positions in universities and industry and has made significant contributions to research on the environmental impacts of energy use, including work on atmospheric diffusion and mass transfer, aerosol dynamics, and chemistry. He is the author of many articles and books on these and related topics. Dr. Hidy received a B.S. in chemistry and chemical engineering from Columbia University, an M.S.E. in chemical engineering from Princeton University, and a D.Eng. in chemical engineering from the Johns Hopkins University.

James R. Katzer (NAE) was manager of strategic planning and program analysis for ExxonMobil Research and Engineering Company, where he was responsible for primary technology-planning and analysis activities and for future-focused technology-planning activities. Prior to that he was vice president, technology, Mobil Oil Corporation, with primary responsibilities for ensuring Mobil's overall technical health, developing forward-looking technology scenarios, identifying and analyzing technology and environmental developments and trends, guiding Mobil's long-term directions on the basis of strategic technical drivers, and identifying future threats and opportunities and recommending strategies to deal with them. Dr. Katzer joined the Central Research Laboratory of the Mobil Oil Corporation in 1981, later becoming manager of process research and technical service and vice president of planning and finance for Mobil Research and Development Corporation. Before joining Mobil he was a professor on the chemical engineering faculty at the University of Delaware and the first director of the Center for Catalytic Science and Technology there. Dr. Katzer has more than 80 publications in technical journals, holds several patents, and co-authored and edited several books. He received a B.S. degree from Iowa State and a Ph.D. in chemical engineering from MIT.

Parker D. Mathusa is a member of the Board of Directors—Research Scientist, New York State Energy Research and Development Authority (NYSERDA). Formerly he was program director, Energy Resources, Transportation and Environmental Research Program, NYSERDA, where he was responsible for establishing research programs and policies required to develop new energy technologies and environmental mitigation measures that could contribute to New York State's energy supply needs, with a focus on renewable energy resources, advanced transportation technologies, and environmental products. Dr. Mathusa's previous positions include service as chief, Utility Research and Demand Management, New York State Public Service Commission, in which he developed a comprehensive R&D program for electric and gas utilities, and engineering positions at Yankee Atomic Electric Company and Bechtel Corporation. He has been involved in the evaluation of a number of emerging energy technologies and associated environmental mitigation measures, including fuel cells, hybrid electric vehicles, and photovoltaic systems, and has published numerous assessments of energy technologies. He has served on numerous advisory panels including federal and state advisory groups. He has a B.S. in physics from the State University of New York at Albany and an M.S. in engineering management from Northeastern University.

Timothy Mount is professor of applied economics and management at Cornell University. His research and teaching interests include econometric modeling and policy analysis relating to the use of fuels and electricity and to their environmental consequences (acid rain, smog, and global warming). Professor Mount is currently conducting research on the restructuring of markets for electricity and the implications for (1) price behavior in auctions for electricity, (2) the rates charged to customers, and (3) investment decisions for maintaining system adequacy. He has spent sabbaticals at the University of New South Wales, Australia, and the London School of Economics and the University of Manchester, United Kingdom. He has a B.S. from Wye College, University of London, and a Ph.D. from the University of California, Berkeley.

Francis J. Murray, Jr., is an energy and environmental consultant, providing strategic policy and market-development guidance on energy and environmental issues for private sector clients. His previous positions include consultant to the Office of Assistant Secretary for Policy and International Affairs, U.S. Department of Energy; chairman of NYSERDA, and commissioner of energy in the New York State Energy Office; deputy secretary and assistant secretary to the governor for energy and environment; and senior legislative counsel/legislative counsel in the New York State Office of Federal Affairs. His experience includes the development and implementation of major energy and environmental initiatives and programs for New York State, including the development of a comprehensive, integrated State Energy Plan that integrated state energy, environmental, and economic development policies in the early 1990s, and policy analysis for the federal government on electric reliability and appliance efficiency standards. He was an environmental policy fellow at the Institute of Ecosystems, Millbrook, New York (1999-2000); director, Scenic Hudson, Inc. (1994-2000); director, the Environmentors Project (Washington, D.C., 1994-2000); and founding member of the Hudson River Greenway Communities Council (1992-1996). He has a B.S.F.S. from the Georgetown University School of Foreign Service and a J.D. degree from the Georgetown University Law Center.

D. Louis Peoples is president and founder of Nyack Management Company, a business consulting and turnaround firm. Formerly he was chief executive officer of Orange and Rockland Utilities in New York State. While at Orange and

Rockland, he was a leader in the deregulation of electric power, serving as chairman of the New York Power Pool and of the Transition Steering Committee to form the New York Independent System Operator. Earlier, he was executive vice president of Madison Gas and Electric Company; senior vice president of RCG/Hagler, Bailly, a consulting company; and vice president of Bechtel Management Consulting Services. Mr. Peoples has also been corporate controller of McGraw Edison Company, director of nuclear licensing at Commonwealth Edison, and training manager at Vermont Yankee Nuclear Power Corporation. He served in the nuclear submarine service in the U.S. Navy. He received a B.S.M.E. from Stanford University and an M.B.A. from Harvard Business School. He is a certified public accountant and a registered professional engineer.

William F. Quinn is founder and president of Argos Utilities LLC. Formerly he was president of Shaw Transmission and Distribution Services, Inc., part of The Shaw Group, where he had responsibility for strategic planning, business development, and the financial viability of the transmission and distribution subsidiaries. Mr. Quinn also sits on the board of directors of Hydro Power Solutions LLC, a joint venture company owned equally by The Shaw Group and Hydro Quebec LTD of Montreal. He also managed The Shaw Group's Structured Transaction Group, where his duties included managing mergers and acquisitions teams, overseeing project development activities, and evaluating investment options. Prior to joining The Shaw Group, Mr. Quinn was responsible for management of the Pacific Gas and Electric (PG&E) National Energy Group's power-asset-development business in North America. Among other projects there, Mr. Quinn directed the 1,200 MW Athens Generating Project, New York's first merchant generating facility and one of the largest gas-fired power plants in the United States. Prior to joining PG&E, he incorporated Meridian Power Corporation, where he was responsible for the marketing, development, financing, and construction of power-generating projects. While at Energy Management, Inc., Mr. Quinn developed several biomass and gas-fired cogeneration projects. He also was project engineer for Badger America, Inc. He has a B.S. in mechanical engineering from the University of Massachusetts and did graduate studies in business administration at Harvard University. He is a registered professional engineer.

Dan W. Reicher is president, New Energy Capital Corporation. He served recently as executive vice president of Northern Power Systems, the nation's oldest renewable energy company. From 1997 to 2001, Mr. Reicher was Assistant Secretary of Energy for Energy Efficiency and Renewable Energy at the U.S. Department of Energy (DOE). As Assistant Secretary, he directed annually more than $1 billion in investments in renewable energy, distributed generation, and energy-efficiency research, development, and deployment.

Prior to that position, Mr. Reicher held other senior management posts in DOE and was also a senior attorney at the Natural Resources Defense Council. He was also co-chair of the U.S. Biomass Research and Development Board, a member of the U.S. delegation to the Climate Change Negotiations, and a member of the board of the government-industry Partnership for a New Generation of Vehicles. Mr. Reicher is also currently co-chair of the advisory board of the American Council on Renewable Energy and a member of the boards of Burrill and Company's Biomaterials and Bioprocessing Venture Fund, the American Council for an Energy Efficient Economy, and the Keystone Center's Energy Program. He has more than 20 years of experience in energy technology, policy, and finance. He holds a B.A. from Dartmouth College and a J.D. from Stanford Law School.

James S. Thorp (NAE) is the Hugh P. and Ethel C. Kelly Professor of Electrical and Computer Engineering and head of the Department of Electrical and Computer Engineering at Virginia Polytechnic Institute and State University. Previously he had been the Charles N. Mellowes Professor in Engineering at Cornell University and director of the Cornell School of Electrical and Computer Engineering. He had also been a faculty intern at the American Electric Power Service Corporation; an Overseas Fellow, Churchill College, Cambridge University; and an Alfred P. Sloan Foundation National Scholar. Dr. Thorp is a fellow of the Institute of Electrical and Electronics Engineers (IEEE) and the editor of the *IEEE Transactions on Power Delivery* for protection systems. Dr. Thorp received the 2001 Power Engineering Society Career Service award. He was a member of the International Advisory Board of the Department of Electrical and Electronic Engineering, Hong Kong University, and a member of the Iowa State Electrical and Computer Engineering External Advisory Board. He has written more than 100 journal articles and many book chapters. He obtained a B.E.E. and Ph.D. from Cornell University.

John A. Tillinghast (NAE) is president of Tillinghast Technology Interests, Inc. Early in his career from 1949 to 1979, he held a number of positions at American Electric Power (AEP) Service Corporation, including executive vice president, engineering and construction, and vice chairman of the board in charge of engineering and construction. Positions that he held subsequent to his employment at AEP include senior vice president and senior technical officer overseeing research and development of Technology Wheelabrator-Frye, Inc.; senior vice president, technology, Signal Advanced Technology Group, The Signal Companies, Inc; and senior vice president, Science Applications International Corporation. His experience and knowledge span a variety of areas, including steam turbines; nuclear energy systems; magnetohydrodynamic power plants; fossil energy power plants; transmission and distribution (T&D) systems; engineering, construction, and operation of electric power production and T&D facilities; restructuring of the utility industry; alternative energy projects; cogeneration including small gas turbines; geothermal plants; life extension of utility facilities; and power marketing. He has served on a number of National Research Council units, including as chairman of the Energy Engineering Board and as a member of the Commission on Engineering and Technical Systems. He is a fellow of the American Society of Mechanical Engineers. He has a B.S. and M.S. in mechanical engineering from Columbia University.

B

Presentations and Committee Meetings

1. COMMITTEE MEETING, THE NATIONAL ACADEMIES, WASHINGTON, D.C. JANUARY 18-19, 2005

Congressional Expectations for the Study
Beth Tritter, Office of Congresswoman Nita M. Lowey, Representative from New York's 18th District

Department of Energy Perspectives: Indian Point Energy Alternatives Study
Philip Overholt, U.S. Department of Energy

Transmission Considerations for the Replacement of Indian Point Generation with Alternate Sources
John Kucek, Oak Ridge National Laboratory

Energy Efficiency and Renewable Energy—Resource Potential in New York State: Summary of Potential Analysis Prepared for the New York State Energy Research and Development Authority (NYSERDA)
Lawrence Pakenas, NYSERDA, and John Plunkett, Optimal Energy, Inc.

Indian Point: What Could Wind Contribute?
Randall Swisher, American Wind Energy Association

Natural Gas Use in Eastern New York: Can the Indian Point Nuclear Facility Be Replaced by Gas-Fired Power Generation?
Harry Vidas, Energy and Environmental Analysis, Inc.

2. COMMITTEE MEETING, CROWNE PLAZA HOTEL, WHITE PLAINS, NEW YORK MARCH 14-16, 2005

Northeast Power Coordinating Council (NPCC) Reliability Criteria, Guides, and Procedures
Philip Fedora, Northeast Power Coordinating Council

New York Power Generation Development Overview
Bill Quinn, Argos Utilities, LLC

ICF Power Market Analysis Capabilities
Juanita Haydel, ICF Consulting

Entergy's Views
Michael R. Kansler, Entergy Nuclear Northeast

Building Transmission Lines
Steve Mitnick, Conjunction LLC

New York State Department of Public Service
Howard Tarler, New York State Department of Public Service

Westchester County Government Views
The Honorable Andrew J. Spano, Office of the Westchester County Executive

Westchester County Legislature Views
The Honorable Michael B. Kaplowitz, Westchester County Board of Legislators

Alternatives to Indian Point
Bruce Biewald, Synapse Energy Economics, Inc; Alex Matthiessen, Riverkeeper; and Fred Zalcman, Pace Law School Energy Project

New York Independent System Operator Views
Garry Brown, New York Independent System Operator (NYISO)

Con Edison Views
Michael Forte, Con Edison

Financing New Electric Generation
Carl Seligson, Economic and Strategic Consultant

3. COMMITTEE MEETING, THE NATIONAL ACADEMIES, WASHINGTON, D.C. MAY 31-JUNE 1, 2005

Integrated Gasification Combined Cycle (IGCC)
N.Z. Shilling, GE

New York State Public Benefits Energy Efficiency Programs
Paul A. DeCotis, New York State Energy Research and Development Authority

4. SITE VISIT, SCHENECTADY, NEW YORK JULY 25-26, 2005

5. CLOSED COMMITTEE MEETING, THE NATIONAL ACADEMIES OCTOBER 17-18, 2005

6. CLOSED COMMITTEE MEETING, THE NATIONAL ACADEMIES NOVEMBER 21-22, 2005

C

Acronyms

AC	alternating current	FERC	Federal Energy Regulatory Commission
AMP	Automatic Mitigation Procedures	FF	fabric filter
AVOC	annual variable operating cost	FGD	flue-gas desulfurization
		FO2	No. 2 (distillate oil)
BWR	boiling water reactor	FO6	No. 6 (residual oil)
C&D	constructing and demolition	GAP	Gap Analysis Program (U.S. Geological Survey)
CAA	Clean Air Act		
CAIR	Clean Air Interstate Rule	GE	General Electric International
CAMR	Clean Air Mercury Rule	GHG	greenhouse gas
CC	combined cycle		
CDW	construction and demolition waste	Hg	mercury
CHP	combined heat and power	HHV	higher heating value
CIPP	Commercial and Industrial Performance Program	HVAC	high-voltage alternating current; or heating, ventilating, air conditioning (Chapter 2 only)
CO	carbon monoxide		
CO_2	carbon dioxide	HVDC	high-voltage direct current
ConEd	Consolidated Edison		
CPU	central processing unit	IC	internal combustion
CRPP	Comprehensive Reliability Planning Process	ICAP	installed capacity
CSP	curtailment service provider	ICR	installed capacity requirement
CT	combustion turbine	IGCC	integrated gasification combined cycle
		IOU	investor owned utility
DC	direct current	IP2	Indian Point Unit 2
DER	distributed energy resource	IP3	Indian Point Unit 3
DG	distributed generation	IPP	independent power producer
DOE	Department of Energy	IRM	installed reserve margin
DR	demand response	ISO-NE	independent system operator-New England
DSM	demand-side management		
		LBMP	locational-based marginal pricing
EE	energy efficiency	LBNL	Lawrence Berkeley National Laboratory
EESP	energy efficiency service provider	LCOE	levelized cost of energy
EIA	Energy Information Administration	LED	light-emitting diode
EOL	end of license	LHV	Lower Hudson Valley
EPA	Environmental Protection Agency	LI	Long Island
ERO	Electric Reliability Organization	LICAP	locational installed capacity
ESP	electrostatic precipitator/precipitation	LIPA	Long Island Power Authority
ETP	Enabling Technologies Program	LNG	liquefied natural gas

APPENDIX C

LOLE	loss-of-load expectation
LSE	load serving entity
MAAC	Mid-Atlantic Area Council (reliability council)
MAPS	Multi-Area Production Simulation
MARS	Multi-Area Reliability Simulation
MDEA	methyl diethanol amine
MIT	Massachusetts Institute of Technology
MSW	municipal solid waste
NAAQS	National Ambient Air Quality Standards
NE	New England
NERC	North American Electric Reliability Council
NG	natural gas
NGCC	natural gas combined cycle
NO_x	nitrogen oxide
NPCC	Northeast Power Coordinating Council
NRC	National Research Council
NREL	National Renewable Energy Laboratory
N-S	north-south
NYC	New York City
NYCA	New York Control Area
NYDEC	New York Department of Environmental Conservation
NYISO	New York Independent System Operator
NYMex	New York Mercantile Exchange
NYPA	New York Power Authority
NYPSC	New York Public Service Commission
NYSERDA	New York State Energy Research and Development Authority
NYSRC	New York State Reliability Council
O_3	ozone
O&M	operation and maintenance
PC	pulverized coal
PJM	Pennsylvania Jersey Maryland (regional transmission organization)
PLRP	Peak Load Reduction Program
PM	particulate matter
PPA	Power Purchase Agreement
PSEG	Public Service Electric and Gas
PUC	public utility commission
PV	photovoltaic, photovoltaics
PWR	pressurized water reactor
REAP	Residential Energy Affordability Program
REPIS	Renewable Plant Information System
RGGI	Regional Greenhouse Gas Initiative
RMR	Reliability-Must-Run
RNA	Reliability Needs Assessment
ROS	rest of state
RPS	Renewable Portfolio Standard
SBC	Systems Benefit Charge
SCPC	supercritical pulverized coal
SCR	Special Case Resource; selective catalytic reduction
SHW	solar hot water
SO_2	sulfur dioxide
SO_x	sulfur oxide
SPDES	State Pollutant Discharge Elimination System
SVC	static VAR compensator
TO	transmission owner
UCAP	unforced capacity
UDR	Unforced Delivery Rights (transmission capacity)
UPNY-SENY	Upstate New York-Southeast New York (transmission interface)
U.S. NRC	U.S. Nuclear Regulatory Commission
VOC	volatile organic compound; variable operating cost
VOLL	value of lost load
WESP	wet electrostatic precipitator/precipitation

Pages 86-178 (Appendixes D through G) are not printed in this volume owing to space limitations. They are on the CD-ROM that contains the full report.